Frank Presbrey, pub Southern Railway (U.S.)

The Empire of the South

Frank Presbrey, pub Southern Railway (U.S.)

The Empire of the South

ISBN/EAN: 9783337170172

Printed in Europe, USA, Canada, Australia, Japan

Cover: Foto ©Andreas Hilbeck / pixelio.de

More available books at **www.hansebooks.com**

EMPIRE of the SOUTH

THIS VOLUME IS DEDICATED TO THE PEOPLE OF THE SOUTH BY THE SOUTHERN RAILWAY COMPANY, WHOSE INTERESTS ARE IDENTICAL WITH THOSE OF THE STATES TRAVERSED BY ITS LINES.

The Empire
of the South.
An Exposition
of the Present
Resources
and Development
of the South.
Frank Presbrey
Published by the
Southern
Railway
Co.
1899

THE advance of the Empire of the South has been one of the grandest and most noteworthy movements in the industrial and commercial history of the world. It has annulled the force of the adage, "Westward the course of empire takes its way," and has destroyed for all time the theory of political economists that emigration follows isothermal lines.

Considered in general, the development of the South in all avenues of human activity has been coincident and parallel to the growth of the country at large. When, however, this great region is considered by itself, or in connection with individual sections of the United States, a basis of comparison is presented which brings out with startling clearness and in incontrovertible figures the majesty and rapidity of its unparalleled progress.

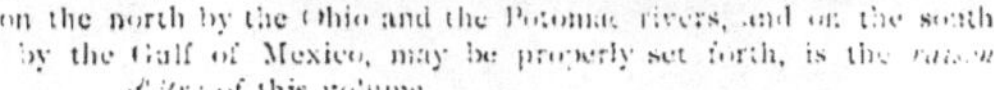

That the record of its growth, and the wholesome and steady development of that portion of the South stretching from the Atlantic on the east to the Mississippi on the west, and bounded on the north by the Ohio and the Potomac rivers, and on the south by the Gulf of Mexico, may be properly set forth, is the *raison d'être* of this volume.

Taken as a whole, the States included in this area form an empire of a half million square miles. It is four times greater than England, Ireland and Scotland, and more than seven times larger than the combined area of the New England States. Within its borders could be placed sixty-four States the size of Massachusetts, and five hundred the size of Rhode Island. It has so generous a supply of natural and material wealth, that, if the balance of the world should be swept out of existence, it could prosper and support itself through the ages to come. Raw materials exist or are successfully grown in every part of the South in such prodigal abundance that transportation from mine and field to factory is a minor item. It has a system of intercommunication and connection with the outside world by water and rail which limits the boundaries of its trade and commerce only as civilization is limited. It has a genial climate and

[illegible] soil, and in all avenues, industrial, com[illegible]l, agricultural, and intellectual, offers its [illegible]itizens, and those [illegible] may in the future [illegible] every advantage and inducement [illegible] in any portion of the United States. [illegible] the South's growth [illegible] comparative figures. Between [illegible] South [illegible] to [illegible] gain of [illegible] per cent., while [illegible] New England and Middle [illegible] combined gained only [illegible] per cent. [illegible] per capita wealth of [illegible] increased [illegible] the same [illegible] 22 per [illegible] while the [illegible] in New [illegible]

there was a gross revenue of 24.1 per cent. on the capital invested in farm interests, while in all other sections of the country the gross revenue was 13.1 per cent. In 1880 the South had $257,244,[illegible] invested in manufacturing. In 1890 she had $657,[illegible], a gain of 150 per cent., while the gain of the entire country was about 121 per cent. The value of the manufactured products of the South in 1880 was $457,454,[illegible]. In 1890 it was $917,[illegible], a gain of [illegible] per cent. In 1880 the factory hands alone in the South received $75,917,[illegible] in wages. In 1890 they received $222,118,[illegible]. In 1880 the South had invested in cotton manufacturing $21,976,000; in 1890, $61,[illegible]; and now about $125,000,000. In 1880 the South had $3,500,000 invested in the cotton-seed oil industry. It has now more than $30,000,000 so invested. The railroad mileage of the South has been increased since 1880 more than twenty-five thousand miles, at a cost in building new roads and in the improvement of old ones of over $1,[illegible]. In 1880 the South made [illegible] tons of pig iron. In [illegible] it made 1,7[illegible] tons. In [illegible] the value of the product was $[illegible]. In 1897 [illegible] value was $[illegible]. In [illegible] the South's output of coal was [illegible] tons. Last year it was 32,872,636 tons, and [illegible] each year [illegible]. The [illegible] of the national [illegible] of the South [illegible] from $[illegible]

of individual deposits from $60,846,500 to $160,875,300 in the same period. These figures are exclusive of savings banks, the deposits in which increased proportionately.

No section is better adapted to the manufacturing industry than the South. It has all needed raw materials in the greatest abundance and of the best quality. Its mineral fields are practically inexhaustible, and they embrace all varieties of ores, and many of them are of surpassing richness. It has coal enough to last for generations, even with the most prodigal use. It [illegible] for refining its ores, and every [illegible] making a first-class quality of pig iron [illegible] can be made in any part of the w[illegible]. It has also [illegible] demonstrated that steel-making [illegible] be [illegible] profitable [illegible]. It has extensive forests of timber, with varieties suited to every kind of wood-working industry, and the forests [illegible] produce immense quantities of tar, pitch, turpentine and rosin.

In building stones it has granite, marble and [illegible], all of excellent quality and in [illegible] quantities. [illegible] for pottery [illegible] earthenware [illegible] clays, [illegible] and other for paint ore.

Besides its larger industries, many smaller ones are constantly being developed by cheap and rapid transportation. Fish and oysters from the South Atlantic and Gulf States find ever-increasing markets in the interior. Early fruits and vegetables are sent in enormous quantities as far north as Canada and the Lakes, and tax the capacity of the railroads in their season, formerly the dullest of the year. Dried and canned fruits are shipped by the trainload, and the Florida orange is crossing the ocean to England after running the Mediterranean fruit off this continent in its season.

It is within bounds to say that, taking into consideration the extent and variety of material, the possible powers of production from the soil and their values, the mineral and forest wealth, the advantages from climatic conditions—temperature, rainfall and length of growing season—the dynamic forces of coal and water power and the advantages given by proximity of interdependent resources, and by geographical position, the natural foundation of the South is four times as great as that of the North. Or, stated in another way, the Southern area, fully developed, is capable of sustaining, in equal prosperity and in greater comfort, four times as large a population as can be sustained in the Northern area under the same conditions.

Much has already been achieved by the South in the creation and accumulation of wealth, and in the appliances for carrying on the work still further. In her towns and cities, her railways and other means

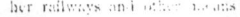

of transportation and the tonnage they carry; in the value of her farms; in mines in operation and their products; in furnaces, mills and factories, and their output; in active capital in the shape of money, credit and organization, in skill in the arts, and in ways and means generally, all considered together, the result of the South's progress has been phenomenal.

With twenty millions of people, and thirty thousand miles of railroad in operation, cotton and other crops of great value, with mountains of coal and ore, with manufactures now large and rapidly growing, with an annual production of iron more than twice as great as that of the United States up to [illegible], and over one-third the world's production up to [illegible], a good start has been made.

Projected through the center of the half million square miles composing that section of the South east of the Mississippi River is a mountainous region of more than one hundred thousand square miles, extending southwestwardly seven hundred miles from the Pennsylvania line into Alabama and Georgia, and having an average width of one hundred and fifty miles.

The northwestern side of this Appalachian region is a continuous, unbroken coal field, embracing forty thousand square miles, and containing forty times the quantity of coal, available to economical mining, which the coal fields of Great Britain held before a pick was struck into the ground. This region is cool and healthy, heavily timbered, and has a soil fairly productive, susceptible of easy improvement, and has the added advantage of a general elevation of two thousand feet above sea level.

Along its southeastern side, from end to end, lies a valley strip of almost equal area, with a general elevation of one thousand feet above sea level, fertile, heavily timbered, the most abundantly and beautifully watered region in the world, rich in a broad and continuous belt of fossil ores along its northwestern rim near the coal fields. At the foot of the mountain ranges, which wall it on the southwestern side, is another bordering belt of brown ores, and between them the marbles, limestones, clays, and other minerals.

Southeast of the valley there is another strip of almost equal area of very high mountainous country, ranging from two thousand to sixty-five hundred feet above sea level, very heavily timbered, full of water power, and rich in slates, fine clays, the crystalline marbles, magnetic and specular (Bessemer) ores, copper, talc, mica, corundum, and other minerals. The wealth of iron matches the wealth of coal. Everywhere, from one end of this region to the other, its interdependent resources, lying in parallel strips, are connected by natural channels worn by innumerable interlacing streams. Upon this field has been made the remarkable development of the South in the past decade, but what has been done has been but the faint scratching on the outcrop. Around this great mound of wealth piled up in the center of the South, forming a natural workshop and a magazine of resources twenty times as great as Great Britain's, lies more than half a million square miles of rich, fertile lands.

"This mountain region alone can furnish permanent employment, when fully developed, for a population twice as great as that of the United States to-day. Standing alone it has combined wealth of soil, climate, minerals, forests and dynamic forces, to sustain and employ a dense population, incomparably greater than the resources of any other region of like area. Its own powers are increased by the varied resources of the Southern and Central Northern States surrounding it. With a population as dense as that of Massachusetts it would contain about twenty-eight millions of people. As dense as that of England and Wales, fifty millions. Compared with Belgium, fifty-three millions. With Saxony, fifty-five millions. The relative inferiority of natural foundation in the countries named will suggest itself to every mind. About it, on all sides, is a country needing the surplus wealth which such population could produce, and able to give back products needed in exchange. The only limit to the growth of wealth, whether in its amount or the rapidity with which it can be created, is the profitable exchange of surplus products between people employed in different work. Distance is the friction—the lost power—of commerce. The nearer to each other that various resources can be worked up for exchange, the smaller the loss. Compact growth is concentrated work. With the proximity of inexhaustible interdependent resources which Nature has given to the South, it has the greatest advantage over the Old World countries, hampered by the long haul of food products and raw materials. They will be less and less competitors as Southern foundations are perfected and industries established. Here, then, is a field for profitable work and investment governed only by the one plain and inflexible law of permanent growth—symmetry. Compared with it, in magnitude of advantages any other field in the world is small."

AGRICULTURE.

The Southland has ever been strong agriculturally, and even before 1860, with only one-third of the population of the United States, it produced more than one-half the farm products of the entire nation. Nature has endowed it with lavish hand in the requisites precedent to successful agrarian development. Its climate is as near perfection as it is possible to attain. Its soil is of such varied constituency that intelligent cultivation makes it possible to produce a variety and wealth of crops unequaled anywhere in the world. It invites the farmer, the planter, stockman, dairyman, truck gardener and florist, and offers the promise of a generous reward for the labor bestowed. There need be no elbowing for room in the South.

and products, and marketable fruits and vegetables. She has a generous soil, a kindly sun, balmy air and plenteous showers, excellent transportation facilities by water and rail, good homes and nearby outside markets.

What else could be needed to make a successful agricultural region. The garden-truck industry, which employs an army of laborers, stretches from the Chesapeake to the Gulf, and has grown to such proportions that it affects commerce and transportation, and the Southern States last season, after supplying the home demand, shipped vegetables and small fruits north to the value of over $15,000,000. The fruit-growing industry has been the counterpart of truck gardening in rapid development, and the few isolated vineyards and orchards of a score of years ago have grown into enormous acreage under profitable cultivation. The Honorable Charles W. Dabney, formerly Assistant Secretary of Agriculture for the United States, and now President of the University of Tennessee, admirably sums up the agricultural advantages of the South in its climatic aspect. After stating that climate is as much a resource of any section as its minerals or soil, he adds:

"We have in the Southern States, owing to the existence of all the important life-zones in broad belts running down the east and up the west side of the Alleghanies, a country capable of

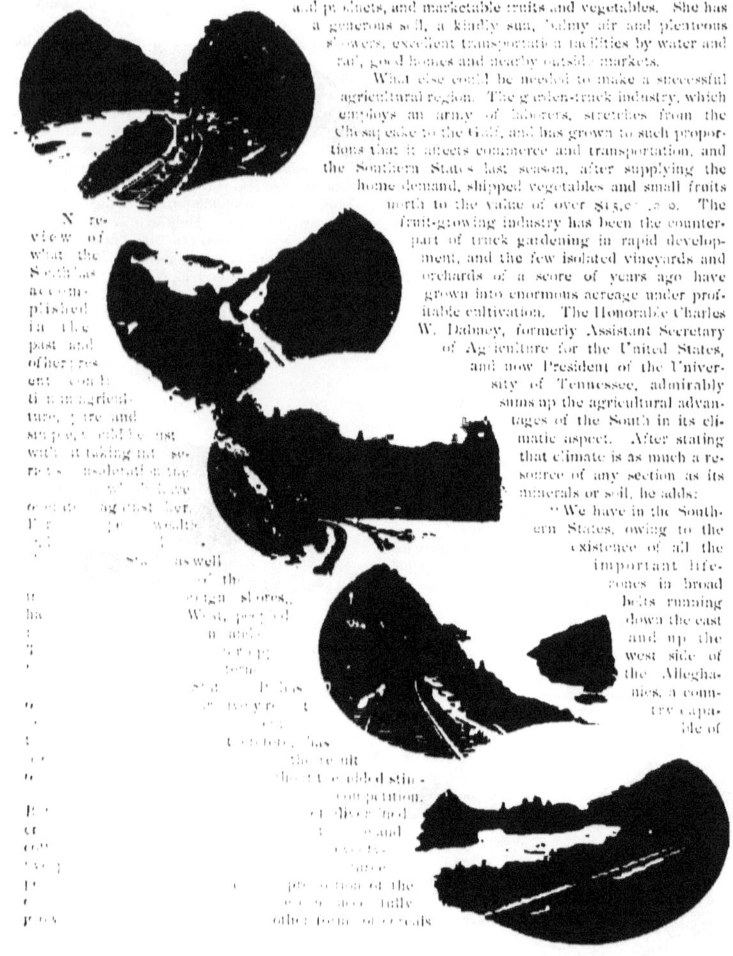

N review of what the South has accomplished in the past and of her present condition in agriculture, ... [illegible]

producing the greatest variety of agricultural and horticultural products—all those, in fact, belonging to the temperate zone, reaching from apples to oranges, from barley to rice.

"The Southern farmer has from sixty to ninety days more in each year in which to work, and during which the sun is working for him, than his Northern countryman.

"While this is true, we have a climate of great equability—not subject to the extremes of either heat or cold. Neither hot waves nor blizzards occur so frequently in the Southeastern States as they do in other sections of our country. The rainfall is as abundant as in the most favored lands on the globe, and is well distributed throughout the growing season, giving sufficient moisture to growing crops even in the warmest months when their demands are greatest. General droughts are rare, and hot winds are not known."

No region in the world offers the large or small farmer better opportunities for a competency than the South, and a study of the statistics of the various States shows the tremendous progress being made in agricultural development.

There are thousands of broad acres along the line of the Southern Railway awaiting intelligent development and cultivation. As demonstrating this fact the most recent authentic statistics give the following figures showing the population per square mile in the countries of the world, compared to the Southern States:

Germany	237	France	[illegible]
Bavaria	[illegible]	Russian Poland	[illegible]
Prussia	[illegible]	Denmark	[illegible]
Baden	[illegible]	Greece	[illegible]
Saxony	[illegible]	Turkey in Europe	[illegible]
Belgium	[illegible]	Russia in Europe (except Poland)	52
Netherlands	[illegible]	United States of America	21
Great Britain and Ireland	[illegible]	The Southern States	[illegible]
Italy	[illegible]		
Austria-Hungary	[illegible]		

The area of the German Empire is 211,108 square miles, a little more than one-fourth as great as that of the South. Its population is 49,421,064. If the South were as densely settled it would have more than 190,000,000 people.

Austria-Hungary has an area of 201,591 square miles, and its population is 41,827,700. With the same number of people to the square mile the South would have 169,000,000.

The area of the united kingdom of Great Britain and Ireland is 120,973 square miles, and its population is now more than 38,000,000. If the South were as densely settled it would have 256,000,000 inhabitants.

The kingdom of Italy embraces an area of 110,665 square miles, and its population is 29,699,000. If the South had as many people to the square mile its inhabitants would number 219,000,000.

The area of the Netherlands is 12,680 square miles; the population is 4,450,870. If the South were as densely populated it would have 287,000,000 people living within its borders.

Belgium has an area of 11,373 square miles, and its population is 6,030,043. If the South had as many people to the square mile as

Belgium its population would be more than 430,000,000.

These figures, however, are likely to be changed during the next decade so far as they relate to the South at least, for the march of emigration is making a wide sweep toward milder climates, and men and women are fleeing from regions of half winter half summer to a more equable zone. They are beginning to discover that it is an immense waste of energy and money to spend so large a proportion of their time in the mere effort to keep warm and comfortable, when they may have that condition for nothing.

To the man of limited means no section holds forth such favorable inducements. Lands are low in price and transportation facilities are of the best. All the grain and vegetable

products that will grow in the West grow much more abundantly in the South, and there is a wide range of products that are indigenous to the South that can only be raised there and cannot be transplanted to the higher latitudes. Rates of living are cheaper than in any other section, because of the mild climate, requiring less fuel, and the greater variety of products available for supplying the necessities of the family. Of the families owning farms, the percentage owning subject to incumbrance, the average incumbrance and the average interest charge are shown in the following table for the whole country and for several Southern States:

FARMS OCCUPIED BY THEIR OWNERS, WHICH ARE INCUMBERED.

	Percentage	Average incumbrance	Average int. charge
United States	[illegible]	$1,224	[illegible]
Alabama	[illegible]	[illegible]	[illegible]
Georgia	[illegible]	[illegible]	[illegible]
Kentucky	[illegible]	[illegible]	71
Mississippi	[illegible]	[illegible]	[illegible]
North Carolina	[illegible]	722	57
South Carolina	[illegible]	[illegible]	[illegible]
Tennessee	[illegible]	[illegible]	[illegible]
Virginia	[illegible]	[illegible]	[illegible]

The logic of the agricultural situation is, therefore, that as a class the Southern farmer has the better end of the financial proposition. The man now living on a rented farm in the overcrowded portions of the North or West has great difficulty in getting a "farm of his own," while if he goes South it is within the power of almost every one to secure a place and be in position to build up and enjoy a home, leaving something for his children to inherit. This is emphasized by the official figures, which show that in the nine seaboard Northern States, with a population of 105 to the square mile, and with 51.81 per cent. of the population urban, there is one pauper for every 550 inhabitants. In the eight seaboard Southern States, with a density of 53, and with 16.03 per cent. of the population urban, there is one pauper for every 1,093.

The vast movements in industrial and mining operations in the South have to a great extent overshadowed the quieter agricultural pursuits, but, nevertheless, tremendous strides were made, as will be seen by the following comparative figures.

Farms ...	[illegible]	[illegible]
Acres under crops	[illegible]	[illegible]
Value of all farm products	[illegible]	[illegible]
Number of live stock	[illegible]	[illegible]
Value	[illegible]	[illegible]

It is little understood among emigrants that the South presents advantages far superior to those of the great West. The climate is much better; the number of towns springing up all over the South bring in their train nearer markets and better prices; the soil and seasons are so admirable that crop failures are rare; the farmer can raise a greater variety of products with the certainty that he can find profitable and convenient markets for them. The small farmer in the South is immensely better situated than one of similar circumstances in the West, and the possibilities in grain-growing in the South were illustrated recently when a South Carolina farmer won the prize offered by the *American Agriculturist* for the largest yield of corn per acre, in competition with the most progressive farmers in every section of nearly every State in the Union.

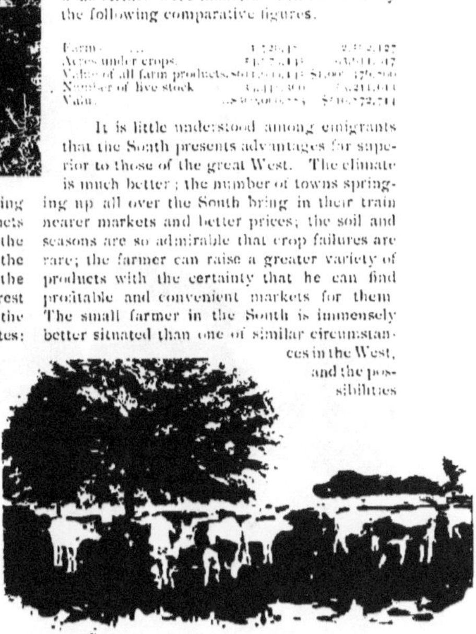

The *Manufacturers' Record*, of Baltimore, has [illegible] that the South's population-supporting [illegible] has scarcely been trenched upon. A[illegible] [illegible]ties, it is possible for the Southern [illegible] a population of upward [illegible] basing the estimate upon [illegible] in Pennsylvania to-day. [illegible] statistics, however, show that [illegible] Southern States, with the excep- [illegible] [illegible]ted to the extent of [illegible] of Massachusetts or [illegible] circumstances, it [illegible] the South [illegible] The South [illegible] an immense [illegible] great [illegible] and. The [illegible] will participate [illegible] largely in [illegible] perity that well [illegible] follow. A [illegible] the [illegible] of [illegible]

business. Nearly every portion of the Southland is well watered and produces nutritious grasses in abundance. Certain sections, as in Virginia, Tennessee, Kentucky and Georgia, have long been famous for the quality of the cattle and horses produced, but as a whole the stock-raising interests of the South are still undeveloped and offer the greatest opportunity for capital and enterprise.

MANUFACTURING

The manufacturing interests of the South are by no means confined [illegible] iron, steel and cotton, although these are entitled to first rank. The practically unlimited water power of the hundreds of streams affords a wealth of opportunities for successful establishments. The great altitude of the mountain regions above the lower [illegible] and, both to the east and west, [illegible] unless [illegible] which may be used either [illegible] now done success- [illegible] the Southern Rail- [illegible] the James, Rappahannock [illegible] Dan River in Virginia; the Cape Fear, Catawba, Broad, Yadkin and Santee in the Carolinas; the Savannah and other rivers [illegible] Georgia; [illegible] Chattahoochee, Coosa, Tallapoosa [illegible] in Alabama; the Tennessee, Holston, [illegible] Pigeon and other rivers in Tennessee; the Kentucky and others in Kentucky, and many other streams, there [illegible] undeveloped sites for the utilization of this enormous power. No other section of the country has such a wealth of opportunities for varied manufacturing at the minimum of cost. Labor is cheap and strikes unknown, power may be had at nominal cost, and raw materials exist in prodigal abundance almost at the door of the factory. To enumerate the variety of products manufactured in the South would be to make a list covering nearly all the needs and uses of mankind, but the great increase in value of manufactured products from $315,924,791 in 1880 to $762,425,300 in 1897 tells an eloquent story of progress. With raw materials close at hand, and the additional advantages of cheap power and competent labor, with a ready home market and unexcelled transportation offered by the Southern Railway to the centers of wholesale foreign and domestic trade, there is no doubt but that the South is admirably adapted to compete successfully with any section of the country.

COTTON.

Cotton has been the great staple of the South for a hundred years, and such it will doubtless continue to be through the coming century. This is simply saying that the causes for cotton's leadership in the nineteenth century will be operative in the twentieth. As these causes are climate and civilization, to doubt their continuance would be like placing a time limit on the law of gravitation. Climate produces the supply of cotton; civilization creates the demand; together they constitute the factors of the leading element of Southern prosperity.

The cotton production of the South for the year ending August 31, 1897, as estimated by Mr. Henry G. Hester, Secretary of the New Orleans Cotton Exchange, was 8,757,964 bales, and the value of the crop $321,924,834. For the past six years the commercial crop has been as follows:

	Bales.	Value.
[illegible]	[illegible]	$[illegible],712
[illegible]	[illegible]	[illegible],512
[illegible]	[illegible],17	[illegible],137
[illegible]	[illegible],251	[illegible],530
[illegible]	[illegible]	[illegible],317
[illegible]	[illegible],757,[illegible]	321,[illegible],834

* The crop grown in [illegible] (marketed in [illegible]) is estimated at [illegible] bales.

The total production for these six years has been 49,102,112 bales, and the value has reached the stupendous aggregate of $1,819,768,072. This does not include the value of the cotton seed, which as at present utilized adds $35,000,000 annually to the resources of the South. The growing number of cotton-seed oil mills, which increased from twenty-five in 1870 to almost three hundred in 1897, is every year changing a constantly enlarging proportion of this potential value into actual value. To every bale of 500 pounds there are generally about 800 pounds of seed, and a ton of this seed yields about thirty-five gallons of oil, valued at forty to fifty cents per gallon. This part of the industry has sprung into existence only in the past ten years, but it is already an enormous business. In 1889 the export of cotton-seed oil amounted to 6,250,000 gallons, and in the next year it reached 14,324,000 gallons. In 1896 over 1,200,000 tons of cotton seed were crushed and about 42,000,000 gallons of oil were obtained. Besides furnishing oil, the cotton seed, after it has been crushed, supplies the cattle with good food in the form of meal and cake, which is claimed to be only a little less nourishing than corn.

Of the world's cotton four-fifths is produced in the Southern States. For the year ending June 30, 1897, they exported 6,176,365 bales, having a value of $230,890,971. Their productive capacity is limited only by demand, and the latter is dependent on the progress of civilization. Every savage won to the ways of light means another consumer of cotton. To be sure, his immediate wants are slight, very likely but a

sack with holes in it for head and arms. But he marks the beginning of a line of shirt wearers. His descendants will want six apiece with starched bosoms. So the demand for cotton grows with enlightenment the world over.

Edward Atkinson has estimated that it would require a crop of fifty million bales to raise the world's standard of consumption to the present standard of the principal nations. At [illegible] rate of increase in the world's consumption there will be [illegible] a demand on the South for sixteen million bales annually, nearly [illegible] present production. At the existing [illegible] capita production, about three hundred [illegible] the cotton States will require a population [illegible] to supply the demand of [illegible] means that the South must add eleven million to her population in the next twenty [illegible] in order to produce the raw cotton that the world will need.

It will be interesting to look for a moment from what the coming years ask the South to do, to what the past years have actually seen her do. In the past will be found an earnest for the future. During the thirty-two years preceding [illegible] the South produced cotton aggregating in value $8,[illegible] 3,391. How vast this sum is can be best shown by comparison. The world's production of gold for five hundred years, from [illegible] to [illegible], was $7,[illegible],0[illegible]0, which is $1,750,[illegible]3,391 less than the value of cotton the South produced in thirty-two years. How like a romance these figures read! What a story they tell of material progress and development! The voyage for the golden fleece seems more probable; but fact is ever stranger than fiction.

In producing this vast aggregate of value the South has barely indicated what she is capable of doing. The United States Department of Agriculture is authority for the statement that, so far as climatic conditions and soil are concerned, there is no limit to the amount of cotton that can be produced by the South until the annual crop is at least ten times what it is at present. If progress be continued in the way of more careful farming, as it doubtless will be, having proved highly profitable, even this estimate will not bound the limit of production. As to the cost of raising cotton, and the many economies experience has taught, much will be found in the succeeding chapters devoted to the various States.

In what has thus far been said, cotton has been considered only as a raw material, but when it leaves the field it has only begun its beneficent mission in the world. From the gin it goes to the railway, the factory, the store, the consumer. Besides the army of cotton pickers, the new crop gives employment to thousands of sailors, captains of steamers and trading vessels, merchants and their clerks, truckmen in the city, and lightermen and longshoremen, and many others. It is estimated that before the cotton reaches the cotton factories it has given employment to nearly 300,000 people in Europe and this country, and that it costs from fifty to sixty millions to harvest a crop. Until recent years the South has contented herself with the production of the raw material. Now she is paying much heed to its manufacture. She has learned that the fabrication of raw materials close to the place of production helps to create that variety in industry which makes a country populous and rich. But the South has not been alone in her learning; the Northern cotton manufacturer has learned that a factory near a cotton field, where he can have cheap coal, cheap labor, and cheap cotton, as he can have in the South, means a decrease in the cost of production and an increase in profits. This knowledge has resulted in the cotton factories of the South increasing from almost nothing forty years ago to 482 to-day, with 3,851,991 spindles, and representing an investment of $125,000,000. Seventy per cent. of these humming spindles that are transforming the South into a mighty industrial center are in the immediate territory traversed by the Southern Railway and its branches, as are 66,561 of the 90,108 looms of the South.

That there is no danger of overdoing the cotton manufacturing business of the South may be seen from the fact that there are in the world about 85,000,000 spindles, representing an investment of about $2,000,000,000, and of this vast industry the United States has a little more than one-fifth in capital invested, or more than $400,000,000, and only about one-fifth of the total number of spindles, or 17,300,000, notwithstanding the fact that the South produces eighty per cent. of the world's cotton crop. It is a noteworthy fact that while the spindles at work in the United States have increased from 10,670,000 in 1880 to 17,300,000 in 1897, the spindles in the South have increased from 584,000 to 3,851,991 in the same period. For one hundred years the South has been raising the cotton, shipping it to New England and to Europe, and permitting the manufacturers to grow rich by turning it into the finished product. As shown, there is practically no limit to the power available for mill purposes, and there is no limit to the cotton available, and as New England can employ 14,000,000 spindles, the continent of Europe 27,000,000 and England 45,000,000, there is no reason why the mills in the South should not continue to multiply for many years to come. Of all the vast wealth of material with which the South has been so abundantly blessed there is no other element, not even iron, equal to cotton in its possibilities of wealth creation for this section. The $[illegible] a year which the cotton crop brings to the South would be trebled if it could be manufactured at home.

The consumption for [illegible] of the [illegible] Southern mills was [illegible] bales, an increase [illegible] bales over the preceding year. This [illegible] the consumption of Southern mills [illegible], the consumption of Northern mills [illegible] almost stationary. While the increase in the number of spindles in Southern mills [illegible] to nearly [illegible] place, the increase in the rest of the country was from [illegible] to 13, [illegible] in the South being about five hundred [illegible] in the whole country outside of [illegible] thirty per cent. In [illegible] the [illegible] fifteenth of the number of [illegible] it has nearly one- [illegible] Nothing could illustrate [illegible]

[illegible]

with the odds in favor of the South, and the superiority of capital with the North. The final outcome is certain. The natural protection of location must in the end triumph over the constant drain necessary to maintain competition under less favorable conditions. This, in fact, is a truism, and the statement is made in no sectional spirit, but as a self-evident proposition."

In the very center of the Northern mill industry, Fall River, Mass., Mr. Joseph Healey, a far sighted New England manufacturer, said recently that in the item of labor cost alone the South had an advantage of twenty-five to forty per cent. over New England. A recent report made by a committee of the Arkwright Club of Boston upon the conditions of Southern competition in cotton manufacturing, and the best practical mode of meeting it, says: "The Southerner finds that with the advantage he possesses he can make these goods at a cost which will allow him to undersell our mills and still leave him a margin of profit which is sufficient to induce the investment of capital. And now, what are these advantages? First, that cotton is conveniently near and that freight on it can be saved; second, that water power is abundant if you care to utilize it, and that coal is cheap if you prefer to run by steam; third, that labor is abundant and cheap and not inclined to organize against the employers; fourth, that the enactment of restrictive labor laws is not liable to trouble manufacturers for many years."

And Edward Atkinson gave the weight of his great authority to the following statement, in a report for the United States Census, showing that New England mills, in cotton manufacture, had an advantage of $3.50 per bale over the mills of Great Britain: "It may be said that this proves too much, and that the cotton spinners of the Southern States will have the same relative advantage over New England. Let this be freely admitted. If Georgia and the Carolinas have twice the advantage over Lancashire that New England now possesses, it will only be the fault of the people of these States if they do not reap the benefit of it." That [illegible] have marked advantages New England no longer denies; that they are reaping the benefit of them all the world knows.

Some of the determining

factors in the movement of the great cotton industry to the South are:

Abundant and cheap water power and coal.

An abundant supply of native American operatives.

Low labor cost because of low cost of living.

Cotton supply immediately at hand.

Cheap and abundant transportation to the markets of the world.

These advantages must inevitably draw the factories to the cotton fields. To say that the South will meet the world's increasing needs, not with bales of cotton, but with bolts of cloth, is merely to say that effect will follow cause.

IRON.

In the making of iron the South has easily the advantage of any other portion of the United States. Her rapid development in this direction has been the phenomenon of the commercial world during the last decade. Not only has she compelled recognition in the markets of this country, but she is now shipping large amounts of foundry iron of the highest quality to Europe, South America and India. Shipments of enormous quantities to Japan, where 5,000 tons recently went in a single week, signify that the limits of her trade are to be confined only to the bounds of civilization itself. When Alabama can undersell English iron four dollars per ton, and make money for the producers, and can underbid Pennsylvania and Ohio furnaces and sell iron under their very eaves, the future of this Southern industry is in a good condition to take care of itself.

The history of iron-making in the South can practically be covered by a span of the last twenty years. The most striking progress has been made during the last decade. Up to 1870 the industry south of the Ohio and Potomac rivers was limited to a few charcoal blast furnaces in Tennessee, Virginia and the Carolinas. The annual output may have reached to 75,000 tons in the best years, or less by twenty-five per cent. than the amount shipped to Europe alone from the South the past year.

A month's output of any of the modern furnaces in Alabama would more than equal the year's production of the best of those earlier plants. In the beginnings of the early-day development Northern ironmasters were loath to believe that any serious competition would result from the introduction of Southern iron on the market. They prophesied that the industry could not last sufficiently long to become a

disturbing element in the market. How much of a factor it has developed into may be gained from the statistics which show that in 1870 the South made six per cent. of the whole country's product of pig iron; in 1880 fourteen per cent; in 1890 sixteen per cent., and in 1896, out of the total aggregate of pig iron produced, the South made 1,850,000 tons, or over twenty per cent. In 1870 the South had $1,516,710 invested in the iron business; in 1897 over $30,000,000, producing an output for the year of 2,250,000 gross tons. Only three European countries make more pig iron than the South—Great Britain, Germany and France. The South is now far in the lead of Austria-Hungary,

Belgium, Russia and Sweden. As showing the advantages of the home market, it may be stated that the consumption of iron in the United States annually is [illegible] pounds for each inhabitant, 280 pounds in Great Britain, 205 pounds in Germany and [illegible] pounds in France.

There were mined in Alabama alone last year over [illegible] tons of iron ore, and this State is now the third in the Union in the production of iron ore, and the fourth in the manufacture of pig iron. Michigan and Minnesota only surpass her ore product, and Pennsylvania, Ohio and Illinois in pig-iron output. Alabama, east Tennessee and Georgia have not only ore but vast beds of coking coal and of limestone in the same localities, and in prodigal quantities. Of late the production of basic pig iron for steel by the open-hearth method has been increasing in Alabama, and so great an impetus has been given to the steel-making industry by the success already attained, that great progress will undoubtedly be made during the next few years.

There have been established a great many foundries, rolling mills, stove works and manufactories along the line of the Southern Railway, using Southern iron exclusively, for while it was formerly supposed that no product could supplant "Scotch pig" for such castings, it has been successfully demonstrated that Alabama iron is its equal in every particular, and the foreign product has practically been driven from the markets.

The Southern foundry trade is a large item in the list of her industries. It has grown rapidly since 188[illegible], especially in gas and water pipe production, plows, and castings for engineering. No statistics of the melting capacity as a whole are obtainable, but good judges place the consumption of pig and scrap iron in these concerns at more than 600,000 tons annually. The largest tonnage goes into pipes and stoves, with agricultural implements and machines second. The most extensive works of this character are at Richmond, Chattanooga, Louisville, Birmingham, and Columbus, Ga.

The furnace "practice" in the South, according to an eminent authority, is equal, for obtaining the best results and economizing expenses, to that of the leading regions of Pennsylvania and Ohio. The superior construction of stacks, more complete utilization of heat, etc., enables Southern masters to make more iron than they made ten years ago, though they now operate fewer plants than they did then.

Southern iron furnaces have been running full time when those of the North and West have been shut down from time to time. The reason for this is that the Southern furnaces, as a rule, are most economically situated as regards their supply of coke, ore and limestone. Northern and Western furnaces buy their ore from the Lakes and their coke from Connellsville or Pocahontas. The Southern furnaces own their coal mines, coke ovens, ore mines and limestone quarries, and themselves mine all their raw material. They pay no profits to coal miners, ore miners or coke makers. They have also another advantage. While the

TOBACCO.

Tradition has it that one stormy night there were gathered at the Mermaid Inn, London, bluff old Ben Jonson, Shakespeare, Beaumont, and a half dozen other genial spirits, when in walked Sir Walter Raleigh, and throwing down on the table some pipes and tobacco, invited them all to smoke and showed them how. Shakespeare is said to have remarked that it was anticipating things a little to smoke in this world, but Jonson, he of ready tongue, after the first pipeful exclaimed: "Tobacco, I do assert without fear of contradiction from yon Avon skylark, is the most soothing sovereign and precious weed that ever our dear mother earth did tender to the use of man. Let him who would contradict that mild but sincere assertion look to his undertaker. Sir Walter, your health!"

From the earliest days of the settlement of the South, tobacco has been one of the main agricultural crops. It was long the chief source of wealth, and for nearly two hundred years the principal currency of the Colonies, and the first loan ever negotiated by the United States Government was made payable in it. Upon tobacco all other values were based, and because of the greater profit in growing it the other agricultural interests were neglected. Prior to the Revolutionary War exports of tobacco had rapidly increased with each year, but during that period its culture in other countries attained considerable proportions, and when peace was restored the foreign market presented a new element of competition, and American tobacco exports have not since that time increased in nearly so great a ratio as before. Its cultivation, however, has extended over all of the Southern States, some growing small and others large quantities. Since 1870 Virginia, which had been up to that time the greatest producer, has ranked second, Kentucky taking the lead.

A cursory review of the history of tobacco-growing presents many points of interest. Probably the first mention of it was made by Columbus on his first voyage, in 1492, when he found the natives using it, and later, on his second voyage, in 1494, Friar Pane, who accompanied him, spoke of its use for both chewing and as snuff. Columbus told further that these natives chewed and smoked an herb having a pungent yet aromatic smell and bitter taste, called cogiaba or cohiba. In 1503 the Spaniards found the natives of Paraguay using it, and in 1519 or 1520 it is mentioned as tobasco. In 1559 some leaves were sent from San Domingo to Europe by Hernandez de Toledo, and a little later Jean Nicot, envoy from the court of France to Portugal, sent to Queen Catherine de Medicis some seed. Through this circumstance it was named Herba Regina, and, in honor of the minister, Nicotina. Still later, in 1565, Sir John Hawkins carried some leaf from Florida to England, and in 1584 a member of Sir Richard Grenville's expedition, which, under the auspices of Sir Walter Raleigh, discovered Virginia in 1585, told of the herb, saying that the natives called it "uppowac," but that in the West Indies the Spaniards called it "tobacco." He goes on to say that the "leaves thereof being dried and brought to powder, they (the natives) used to take the fume or smoke thereof, by sucking it through pipes made of clay into their stomache and head."

In 1610 the first secretary of the Virginia colony wrote: "Here is a great store of tobacco which the savages call apooke, howbeit, it is not of the best kind; it is but poor and weak, and of a byting taste. . . . . The savages here dry the leaves of the apooke over the fire, and sometimes in the sun, and crumble it to powder stalks, leaves and all—taking the same in pipes of earth, which they very ingeniously can make." In 1585, when Sir Richard Grenville returned to England, he carried with him both pipes and ... who was ... first ... England

Kentucky leads all the Southern States in the total amount of its production, with Virginia, North Carolina and Tennessee next in order named.

... crop of the country ... produced on ... Of the ... were grown ... Southern ... average ... almost

reason of the distance from places of consumption, practically for the most part out of the market, and are being decimated by reckless lumbering and fires so rapidly that even in amount they will soon be less than the Southern resources. Prof. B. E. Fernow, Chief of the Forestry Division of the U. S. Geological Survey, says the South contains not only the largest amount and the greatest variety of hard woods, but it also contains in the greatest abundance and perfection that most important class of timber which furnishes three-quarters of our lumber consumption—the pine and its coniferous substitutes like the cypress, cedar, spruce and hemlock. The importance of this fact will appear more strikingly in a few years, when the white pine supplies of the Northern States will have been decimated and brought to a subordinate condition. At present, of the nearly thirty billion feet of pine and other coniferous lumber used in the United States, the Northern States furnish the bulk, the Southern States a little over one-quarter. But presently the white pine of the North, which now reaches an annual output of eight to twelve billion feet of material, will gradually decrease, in fact it has already begun to decrease, and in the same proportion the output of Southern pine must increase. Northern lumbermen are investing in Southern pine rapidly, and in a few years the center of lumber production will be found south of the Ohio and Potomac rivers.

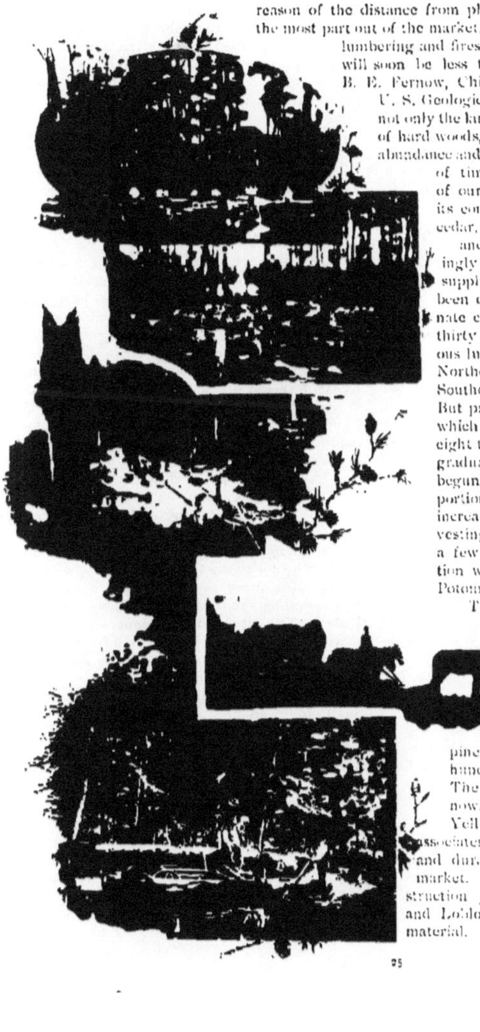

The Southern pine belt, stretching with a width varying from 100 to 200 miles along the Atlantic and Gulf coasts, and containing nearly one hundred and fifty million acres, contains not less probably than twenty-five million acres of unculled virgin pine, and altogether probably over two hundred billion feet of standing pine. The quality of this pine is world renowned, especially that of the Longleaf, Yellow or Georgia varieties, and their associates the Cuban pines, which for strength and durability excel all other pines of the market. It is the material for heavy construction *par excellence*, while the Shortleaf and Loblolly pines furnish excellent finishing material.

In addition to the wood, these pineries furnish annually from seven to eight million dollars' worth of naval stores, rosin and spirits of turpentine; and, as investigations of the Division of Forestry of the United States Agricultural Department have lately shown, without impairing the value of the wood.

A most excellent and pleasing substitute for white pine in house finishing is furnished by the bald cypress, the Big Tree of the South, which haunts the swamps along the rivers. Its lasting qualities in contact with the soil, or in the shape of shingles on a roof, have long given it foremost rank among durable woods.

The mountains of Georgia, Tennessee and North Carolina contain considerable though scattered areas of the northern conifers, white pine and spruce, while hemlock skirts the mountain streams. But the features which have made these mountain forests famous are the big tulip trees and magnificent development of oak and other hard woods. Trees of diameters over five and six feet, and one hundred feet to the first limb, are [illegible] common. This large-sized material, to be sure, is not found spread over the whole mountain range, but [illegible] in coves and small areas here and there. Being to a degree secluded and [illegible] means of transportation, it has been waiting for enterprise and development, [illegible] justify the extension of railroads into [illegible]

The [illegible] of Kentucky and Tennessee and the [illegible] of Georgia, Alabama and M[illegible] in this wonderful hard-[illegible] especially along the many river [illegible] M[illegible] and Y[illegible] delta has con-[illegible] areas of hard [illegible] rich in oaks. The [illegible] to which Northern [illegible]

M[illegible] species may be [illegible] Southern hard wood [illegible] are of high [illegible] will be [illegible]

excellent dimensions and quality; red gum, vying in size with the tulip trees, only a few years ago despised, now a well-established article; chestnut, beech, elm and hackberry, not to forget black walnut and cherry, of which the South still claims available supplies.

If the center of pine lumber production is soon to be in the South (766,420,000 feet were cut in 1896 in the States reached by the Southern Railway), the center of hard-wood lumber production has for some time been located there.

STONE AND MINERALS.

The South has an opulence of building material both above and below ground. The forests with their giant trunks for joist and rafter find a complement in the quarries of granite, marble and other building stone for foundation, wall and ornamentation. Without a single ship from Tarshish or a cedar from Lebanon the South could duplicate the temple of Solomon, drawing every needed material from within her own rich borders, even to the gold for the candle-sticks and the precious gems to sparkle from the altar.

The marbles of East Tennessee are second only to those of Carrara. There are over two hundred varieties of them, each distinct from the others. The exquisite tints and variegated beauty of one variety are the admiration of every visitor to the Capitol and the new Congressional Library at Washington, and other State and national buildings throughout the Union. The output of the Tennessee quarries reaches into millions of dollars. In North Carolina and every State reached by the Southern Railway there is found building stone of the highest quality and in an abundance that makes quarrying profitable.

In several of these States, moreover, there are precious metals in paying amounts, notably in Virginia, the Carolinas and Georgia. Surprising as it may seem to those who have come to look upon the far West and the far North as the only gold regions, the South has produced over $[illegible] worth of the yellow metal, more than [illegible] having come from a single North Carolina mine. The Government mint report that from the beginning of the century to the present time the amount of gold produced in Virginia has been $3,[illegible]3,000;

North Carolina, $21,700,000; South Carolina, $3,55[illegible]; Georgia, $1[illegible]; Alabama, $1[illegible], and Tennessee, $16[illegible].

EDUCATION

Aside from developing her material interests the people of the South have always taken a most earnest interest in the things which make for better citizenship, notably in the direction of the education of her young. During the past thirty years, five hundred and thirty million dollars have, according to the most competent

estimates, been expended in the South in the building and maintenance of the schools and colleges. There is not a community in all the South where there are not ample common [illegible] of facilities, and in all the States there are universities of high rank, and numerous denominational and non-sectarian colleges, seminaries and academies. Many technical and industrial schools have been established and are in flourishing condition, and education for the [illegible] as well as the [illegible] is provided.

The So[illegible] has [illegible] teachers at work, [illegible] many [illegible] and [illegible] 4,[illegible] children in attendance at public and private schools. It is spending [illegible] a year for public education, or nearly four times as much as it did [illegible] years ago.

[illegible]

PORTS.

The Southern Railway meets the sea at Norfolk, Va., where it has extensive wharf facilities at Pinner's Point and West Point, Va., and at Brunswick, Ga. At each of these places it transfers to the coastwise and foreign-bound ships the products in raw and finished materials from the mine and mill, and enormous quantities of cotton, grain and fruit.

Baltimore, with its great maritime interests, is also brought into touch with the Southern Railway system by the Baltimore, Norfolk and Richmond Steamboat Company, which is owned by the railway, and which operates a line of high-class steamers between Baltimore, West Point and Norfolk. At Richmond the Southern Railway connects with the various river lines and the Old Dominion Line for New York. To the westward its water gateways are at Cincinnati and Louisville on the Ohio River, and Memphis, Tenn., and Greenville, Miss., on the Mississippi.

The Eastern harbors are much nearer the wheat, grain and meat producing districts than any of the North Atlantic ports. St. Louis, for instance, is [illegible] miles, air line distance, from New York, 7[illegible] from Norfolk and West Point, Va., and [illegible] from Brunswick, Ga. In the adjustment of future transportation problems these distances will be leading factors, the Southern Railway having the additional advantage of never being blocked by snow or ice. Already the exportation to [illegible] of Western grain and meat products has grown to impres[illegible] [illegible] through [illegible] Brunswick, and it has [illegible] established [illegible] there is [illegible]

UPPER WATERS OF THE RICHLAND RIVER—LAND OF THE SKY

The growth of shipping to and from the Southern ports has been the marvel of the maritime world. One port showed a gain in exports of breadstuffs alone in the year 1897 over 1896 of 137 per cent , another 171 per cent. The latest figures of the statistical department of the United States Treasury Department show that the increase of exports of this class at the four chief Northern ports in 1897 over 1896 amounted to $7,019,510, or 74 per cent., while the increase from the four chief Southern ports was $7,944,151, or 163 per cent.

While these figures cover but one line of goods, the increase in other products was equally great, the variety of exports being greater with each succeeding year.

MOUNTAINS.

No other mountain region is to be in any way compared with the magnificent section in the western portion of North Carolina and eastern Tennessee poetically called "The Land of the Sky." Here are forty-three distinct peaks higher than Mount Washington, eighty which are more than five thousand feet in altitude, and countless scores exceeding four thousand. From one of these "fortresses of nature" seven different States may be seen and the eye may bring within its span fifty thousand square miles, a wild billowy area where range after range of forest-clad peaks follow each other as waves chase up a beach.

The Appalachians, as the various mountain ranges are called which constitute the great eastern border mountains of North America, and reach their highest altitudes in western North Carolina and eastern Tennessee, originated ages ago in processes of upheaval and were completed just after the close of the carboniferous period. They are composed of great masses of sedimentary rock which once lay beneath the sea. Their history is a long one, and to the geologist and physiographer one of great interest. The arrangement of narrow valleys and linear ridges presented in this mountain system is such as to make a type of topography which is nowhere else on earth so characteristically and extensively developed. The Appalachians have a generally southwesterly and northeasterly trend for over one thousand miles, and extend from southern New York through the States of Pennsylvania, Maryland, Virginia, Tennessee, North Carolina and South Carolina, terminating in northern central Alabama.

In Pennsylvania the range reaches an elevation of 2,000 feet above the sea, or 1,000 to 1,500 feet above the adjacent Cumberland Valley. At Harper's Ferry the historic eminences of Maryland Heights and Loudon Heights overlook the Potomac at an elevation of 800 feet. Southward through Virginia, however, the range becomes broader and higher. Forty-five miles below the Potomac is Mount Marshall, 3,150 feet high, and a short distance farther, near Luray, Stony Man and Hawk's Bill, 4,031 and 4,066 feet, respectively. These are the highest summits of the Blue Ridge north of North Carolina.

The Piedmont Plain in Virginia, which the main line of the Southern Railway traverses, extends along the southeastern base of the Appalachian Mountains. Its surface has a gentle eastward slope from an altitude of about 1,000 feet at the western edge to 250 or 300 feet on the east, where it merges into the Coastal Plain.

Through Virginia, North and South Carolina and part of Georgia the western limit of the Piedmont Plain is along an irregular line in which the gentle slope of the etched plain changes to the steeper slopes of the Blue Ridge.

The most striking characteristic of this range is the great difference in slope of its opposite sides. The streams heading in the gaps upon the divide flow westward in broad, smoothly rounded and drift-filled valleys for miles before entering the narrow rock-cut gorges of their lower courses. Those flowing eastward, on the other hand, plunge immediately downward in a series of cascades, falling several thousand feet in a distance of a few miles. They have no valleys, only V-shaped gorges, until they reach nearly to the level of the Piedmont Plain. The difference in slope is admirably shown on the line of the Southern Railway from Salisbury, N. C., to Asheville. From Asheville [illegible] the road ascends the [illegible] the Swannanoa, with [illegible] grade, making [illegible] gap. Passing the [illegible] summit [illegible] the head [illegible] the Catawba by an [illegible] series of loops, winding back and forth [illegible] [illegible]

[illegible] most of the peaks [illegible] nearly the same [illegible] and are [illegible] waves on a choppy [illegible] and less dis[illegible] outlines are barely distinguish[illegible] at the horizon. The [illegible] are generally hidden from [illegible] an occasional clearing on [illegible] and the grassy "balds" on [illegible] higher domes, the whole region [illegible] be covered with a forest mantle.

Only rarely does a ledge of naked rock appear through the vegetation, so that the slopes are smoothed and softened and the landscape lacks the rugged character of unforested mountain regions. The atmospheric effects also tend to produce the same result. The blue haze, which is almost never absent from this region, and which is recognized in the names of both the Blue Ridge and the Great Smoky Mountains, softens the details of objects comparatively near at hand, and gives the effect of great distance to peaks but a few miles away. By reason of this atmospheric effect these mountains of only moderate altitude often afford more impressive views than heights and distances two or three times as great in the clear air of the West.

A very large number of the interior summits reach altitudes between 4,000 and 5,000 feet, and a few are over 6,000. The Black Mountains, a few miles north of Asheville, contain the highest peaks in the Appalachian Mountains. Mount Mitchell, altitude 6,711 feet, is the highest point east of the Mississippi, being 425 feet higher than Mount Washington.

RESORTS AND CLIMATE.

In the line of health and pleasure resorts the South is particularly fortunate, both as to the great number and to their wide variety. Many people, especially those living in the North and West, think of the South only as a place to be visited in the winter season. As a matter of fact, there is no region in America which holds out greater inducements to the tourist at any season of the year, both as to scenic and climatic advantages, than the "Land of the Sky" in western North Carolina and eastern Tennessee. The average summer temperature at the mountain resorts in this region is several degrees lower than in either the White Mountains or in the Catskills. This is accounted for by its altitude, which ranges from 2,200 to 6,700 feet above sea level. In winter this same section attracts thousands of visitors from the North because of its wonderful freedom from dampness. So remarkable is this climatic characteristic that the United States Government has issued special scientific bulletins in explanation.

In summer this fair "Land of the Sky," of which Asheville is the commercial and social center, is one of the most enjoyable regions in all the world for recreation and rest. Of late years it has become what Switzerland is to Europe—an international playground.

But the all-the-year-round pleasure and health resorts of the South are by no means limited to Asheville, Hot Springs and neighboring places in North Carolina. There is Lookout Mountain, as well as the Tate Springs and numerous others in Tennessee, the Lithia Springs and Brunswick, Cumberland Island and St. Simon's Island in Georgia, the ever popular Old Point Comfort, Virginia Beach and others in Virginia, all of which are equally enjoyable to the visitor, whether his sojourn be during the winter or the summer season.

Those resorts which are chiefly enjoyable in winter are of world-wide reputation.

Augusta, Ga., and Aiken, S. C., since their attractions, both health-giving and for recreation, have become known, have grown into

[illegible] At each [illegible]

and Palm Beach on the east coast, and Tampa, Punta Gorda, Bellair and Tarpon Springs on [illegible] is familiar. The re[illegible] [illegible] mainly in summer include the [illegible] Flat Rock, Tryon, Hay[illegible] [illegible] Springs, and Linnville, N[illegible] [illegible] Greenville and [illegible] Warm Springs, [illegible] Mt. Airy, [illegible] Gaines[illegible] [illegible] Mountain, [illegible] Springs, [illegible] Spring, [illegible] Alle[illegible] [illegible] Anniston, and Glen A[illegible] [illegible] Monte Sano, Alabama.

[illegible] great has [illegible] in [illegible] development of Southern pleasure places. A[illegible] resort hotels of the South [illegible] accustomed to the refinements [illegible] would be content to sojourn, [illegible] the fingers of one hand. [illegible] two score reached by [illegible] at which the enter[illegible] [illegible]

S[illegible] are confessedly more beau[illegible] than any others of [illegible] America. As a rule the [illegible] him in all portions [illegible] managed and thoroughly [illegible] which the entertainment [illegible] to meet the approval of the [illegible]

[illegible] apprehension regard[illegible] [illegible] in the South, the table [illegible] giving the [illegible] U. S. W[illegible] [illegible], will [illegible] W[illegible] South and the rest of [illegible]

SPORT.

The opportunities for all varieties of shooting and fishing in the South are most excellent, and the seasons are so extended that out-of-door life is enjoyable during the entire winter.

Virginia and North Carolina have long been favorite regions for quail-shooting, and these swift-winged denizens of woodland and stubblefield are undoubtedly more abundant in these two States than anywhere else north or south. They are to be found, however, in satisfactory numbers in all of the Southern States. In South Carolina, Georgia, Florida, Alabama, Tennessee and Mississippi they are usually very plentiful, but in the more southern regions they do not attain the size, nor are they as strong and swift of flight, as in North Carolina, Virginia and Tennessee.

The great salt-water bays and marshes of the coast of North Carolina, South Carolina and Georgia teem with ducks and geese, while brant and swan may be killed in large numbers in season. There is most excellent sport of this class also to be had on many of the streams in Alabama and Mississippi.

The smaller water birds, such as rail, reed birds, snipe and plover, are plentiful all along the coast from Norfolk to Florida, and the sportsman will find especially good shooting of this class in the neighborhood of Morehead City, N. C., and Brunswick, Ga.

Woodcock are plentiful in many places in Virginia, North Carolina, Tennessee and Kentucky, and wild turkeys are found in all of the Southern States, being particularly abundant in Florida.

While Virginia has long been a favorite resort for deer hunters, each of the other States offers good shooting. In Georgia, Florida, Alabama and Mississippi deer are especially plentiful, and are killed each season in such numbers as to astonish the average sportsman of the North. There are too many sections where good shooting may be had to allow of enumeration.

In the mountain regions of western North Carolina and eastern Tennessee many black bears are killed each winter by the hardy sportsmen who have the courage to undertake the work.

The mountain streams offer the best of brook-trout fishing, and in several of those in North Carolina which have been systematically stocked the large rainbow trout are taken by the skillful angler in satisfactory numbers. Black bass are found in great numbers in Virginia, the Carolinas, Tennessee and Kentucky.

The region round about Brunswick, Ga., is the best on the Atlantic coast for salt-water fishing, an infinite variety of sea fish being taken in the nearby waters.

There are many other places where most excellent luck will attend the sportsman, notably the famous resorts on the Gulf Coast and Florida. No section of the country is comparable to the South to-day in the great variety and quantity of game. There are excellent game laws in nearly all the States, and visiting sportsmen are always welcome.

FINIS.

In the foregoing pages there has [illegible] presented in a general way a record of the [illegible] gress which the Southland has been making in the [illegible] of material devel[illegible]ent. A more detailed treatment will be found in the chapters upon the various States.

The [illegible] current [illegible] sperity and progress [illegible] it is sweeping [illegible] the South is [illegible] with every swing [illegible] ture. Every ship that leaves her ports [illegible] foreign [illegible] is heavier laden, every mile of railroad [illegible] is bearing [illegible] greater train [illegible] her harvests [illegible] The hum of her sp[illegible] the [illegible] mining [illegible] in industrial [illegible] into millions [illegible] lifting [illegible] their [illegible] prosperity.

[illegible] the [illegible] Mississippi [illegible] the Gulf [illegible] evolution, more mighty in its significance, more powerful in its influence than any the world has ever [illegible] known, is being wrought. The pulse beatings of this awakening are felt in every artery of trade and commerce in this and foreign lands. Sections in the North where generations have succeeded each other in controlling the markets in cotton goods, confess their inability to meet the more practical conditions of manufacturing in the South. Her people are in earnest, and have set their faces toward the goal of prosperity with a determination kindled by hope and augmented by success already attained.

The future of the Southland? By every [illegible] of material riches it should and will be [illegible] brilliant than that of any other section [illegible] Union. Her gracious smile awaits the [illegible] incoming immigration. Her broad and [illegible] acres, her fertile mountain and hillside [illegible], her rich valleys and crystal streams, her mines of coal and iron, her untouched forests, [illegible] majestic, all pulsate with quickened [illegible] stretch forth the hand of welcome and the [illegible] promise of prosperity.

THE city of Washington, with its massive and historic national buildings, its miles of smooth avenues and countless beautiful and stately residences, its scores of elm-shaded parks and its picturesque suburbs, easily maintains its proud distinction of being the [illegible] attractive and alluring of our American cities.

C[illegible]lly or industrially considered, it is not [illegible], but it is a great capital, and as such will ever be held dear to the hearts of all true Americans.

[illegible]harms of Washington unfold themselves readily to any one who yields

THE [illegible]

to their [illegible] There is [illegible] [illegible] The [illegible] [illegible] the [illegible], and the [illegible] [illegible]

capital. Its very life, commercially and socially, is so closely interwoven with governmental affairs that all else is subverted and appears insignificant. The visitor finds within it a touch of Paris, a suggestion of Berlin, and definite impressions of various American cities. It is at once cosmopolitan and provincial, and its aspects, like its population, are largely changeable.

The physical transformation of Washington from the miserable apology of a town that it was in the sixties to the magnificent city of to-day has been but little short of marvelous. The relics of the earlier days have almost wholly disappeared, and there have grown up in their stead many substantial and modern structures which bespeak the touch of wealth and refinement. This is especially true of the residential section, which for variety of [illegible] and suggestions of refinement [illegible] most favorably with any city on this continent.

The chief center of interest in Washington [illegible] Capitol, and it is impressive from whatever [illegible] at whatever hour it may be viewed. [illegible] the world is its better in beautiful

CONGRESSIONAL LIBRARY

symmetry or majestic dignity. Its nearby neighbor, the newly completed National Library building, is acknowledged to be without a peer on either side of the Atlantic in architectural effect or decoration. At the other end of Pennsylvania Avenue, which is the great main artery of Washington, stands the Treasury Building, impressive beyond description in the very simplicity of its classic façade. Beyond the Treasury, and surrounded by widespreading elms and velvety lawns, is the historic White House, about which cluster a myriad of our nation's fondest memories. From its rear porch one may look across a mile of beautiful mall, stretching away to the very edge of the placid Potomac, and see silhouetted against the southern sky the graceful lines of the towering Washington Monument. Near the White House is the magnificent granite structure occupied by the War, State and Navy Departments, and which will well repay the visitor for the time spent in visiting them. Southeast of the Monument is the huge building known as the Bureau of Engraving and Printing, in which the paper money of the Government is made. From the Potomac to the Capitol is a beautiful stretch of park, in which are located, amid a forest of stately trees and acres of beautiful lawns, the Smithsonian Institution, the National Museum, the Fish Commission Building, and at the eastern end of the park, and almost under the shadows of the noble Capitol, are the Government greenhouses and conservatories, surrounded during the summer season by a wilderness of beautiful flowers and rare plants and shrubs.

Washington might well be called a city of parks, for in addition to the nearly two hundred circles and triangular reservations, where the wide avenues named for the States cross the streets diagonally, there are several large and beautiful squares rich in foliage, statues and ornamental flower beds. Outside the limits of the city proper there is an immense park area, including the Soldiers' Home grounds of three hundred acres and the National Rock Creek Park, which is nearly seven miles from end to end, and includes the most picturesque portions of the Rock Creek Valley.

Turning from the beauties of nature in and about Washington to the beautiful in art will lead the visitor to the handsome new Corcoran Art Gallery, which embraces one of the finest collections of paintings in the country. It is one of the most frequented places in the city, and is open to the public daily.

Situated on the south side of Pennsylvania Avenue, at the corner of Thirteenth Street, is the large and imposing administration building of the Southern Railway. As Washington is the gateway to the Southland from the North and East, there is a sentimental as well as business justification for locating here the headquarters of this, the greatest and most comprehensive transportation company in the South.

THE traveler of to-day, surrounded by all the luxuries which the very mention of a vestibule limited train implies, and engrossed in the problems of modern business, will not, in any probability, as he speeds across the Old Dominion State, dwell upon the fast-fading legends and historical heirlooms of her Colonial days. Yet no other State is so rich in all that is interwoven with the early history of America and our nation as Virginia. She was the cradle of liberty, the natal place of several of our early Presidents, and also of those great leaders who hewed out the strong foundation timbers of our national structure. To her shores came the earliest colonists from England, and here it was that the first settlements took foot-hold. So closely is the history of this great State intertwined with that of the nation that to tear them apart would be to destroy the fabric of both.

There have been six epochs in the history of Virginia which mark as milestones the various periods of her existence. Each one stands to a certain well-defined degree apart from the others; each has produced its leaders and has exerted its far-reaching influence upon the growth and development of the nation. First comes the period of settlement, to recite the history of which is to retell the story of the fortitude and struggles of the Jamestown colony. Following this are the Colonial days, in which there were duplicated in the Old Dominion the great estates, the princely entertainment and the aristocratic country-house life and the politics of England. Next in turn is the Revolutionary period, which gave us Washington, Jefferson, Henry and a host of other patriots. Then the era of Statehood. Subsequently, her withdrawal from the Union, and her vast influence on her sister States in the South, and to-day the progressive and intellectual modern commonwealth, resonant with the hum of the factory and workshop, rich in agricultural resources, and resplendent in achievement in all lines of human activity.

Virginia has twice as many grand divisions as had the ancient Gaul of which Cæsar wrote. These are the Tidewater, the Midland, the Piedmont, the Blue Ridge, the Valley of Virginia, and Appalachia, or the mountain country. These divisions not only succeed each other geographically, beginning on the east, but they differ in relief, occupying different levels above the sea. From the Atlantic to the west they rise like a natural stairway, the topmost step in the mountain [illegible] feet in elevation.

Speaking broadly, the State may be divided into a lowland and a highland country. Its [illegible] part, over [illegible] square miles, or rather more than half of the whole State, has the aspect of a broadly undulating plain, that, with but few marked variations of relief, rises from the sea to from [illegible] to [illegible] feet above the [illegible]. The northwestern portion, a part of the region widely known as the Atlantic Highlands, is one composed of approximately parallel mountain ranges, running entirely across the State from northeast to southwest, separated by nearly parallel valleys, some of them wide and others narrow, varying in breadth from a half mile to twenty-five miles, the whole surface presenting all the varieties of relief peculiar to the Appalachian country between the altitude levels of [illegible] and [illegible] feet. Speaking more [illegible], however, the State is naturally divided into the six grand divisions above [illegible]. In [illegible], soil and product, as well as in elevation, these divisions vary. Taken altogether they offer an abundance and variety of

who included in their membership John Marshall, afterward Chief Justice of the United States.

Twenty miles south of Culpepper the train rolls across the historic Rapidan, and shortly after Orange is passed, [illegible] which [illegible] Montpelier, where a glimpse may be [illegible] in passing of the home of James Madison, the fourth President. It is a beautiful region, all of it, from Alexandria to Charlottesville and then on south, passing North Garden, Amherst and Monroe to Lynchburg. Leaving this prosperous city the road follows its southwesterly course to Danville, passing on the way Lawyer's Road, with its nearby [illegible], [illegible] and [illegible]. [illegible] this portion of the route that the Blue Ridge Mountains begin to [illegible] against the western [illegible] [illegible] lovely panorama [illegible]

Beginning [illegible] which a French savant once called [illegible] it will be worth [illegible] [illegible] detail to [illegible] [illegible] the [illegible] [illegible] [illegible] [illegible] [illegible] [illegible] [illegible] crop alone.

[illegible]

Every variety of fruit which will grow in the temperate zone flourishes in the Piedmont region. The sunny slopes of the mountains have a peculiarly light soil, kept constantly fertile by the decomposition of rocks [illegible] potash, and perennially moist by numerous springs. This soil is, therefore, admirably adapted to apples, and one of the most famous kind—the Albemarle pippin—has been brought to its highest perfection here. It is the favorite in foreign markets, and usually sells at three dollars or more a barrel on the trees, the buyer furnishing the barrels and doing the picking.

Next in importance to the apple comes the grape in line of fruit. A peculiarity noted in the most favored claret-producing vineyards of France is the large admixture of iron in the soil. This is the characteristic of much of the soil of this section. The soil and climatic conditions of Virginia, when compared with the grape districts of Germany and France, present many striking similarities. The average ranges of the thermometer of this section and of those at Bordeaux and other vine-growing sections of Germany and France are very close together. In the rolling, hilly country, with its calcareous loam, or gravelly, loose soil, with a rocky subsoil, facilitating self-drainage, with exemption from heavy spring frosts and early frosts in autumn, with rarely any excess of rainfall in the maturing months of June, July, August and September, are the most favored conditions for the vine. This is shown, as would be supposed, by the luxuriant growth and fine quality of the native uncultivated grape. A wine-making industry of no small volume has in consequence prospered at several prominent [illegible] of the region, notably at Charlottesville. The products of the [illegible] are the pure fermented grape juice. [illegible] allowed to [illegible] the "bouquet" that age alone can give, they [illegible] comparison with some of the noted wines of Europe. The Piedmont region [illegible] properly called the "fruit belt" of Virginia, and its adaptability to fruits and vines, when properly [illegible], will [illegible] [illegible] [illegible] [illegible] plums, cherries, apples, pears and [illegible] [illegible] [illegible] [illegible]

All [illegible] Virginia, [illegible] [illegible] and dairying [illegible] among the [illegible] for deep farming [illegible]

The Great Valley of Virginia, strange as it may appear, although so peculiarly adapted to grazing and agr[illegible] by [illegible] of its lands, still [illegible] by park-like [illegible] and other hard-wood [illegible] a character as to [illegible] that the makers of agricul[illegible] [illegible] of timber. [illegible] Virginia is peculiarly [illegible] valleys [illegible] lower [illegible] occupied [illegible] a score [illegible] kinds of valuable timber [illegible] Spanish and [illegible] poplars, ash, hickory [illegible] sought for timber trees [illegible] the mountain ranges are [illegible] pine and other [illegible] red and chestnut [illegible] white and black [illegible] timber trees. The [illegible] of abundant [illegible] railway ties, telegraph [illegible] for paper pulp and for [illegible] the kinds demanded [illegible] and structural purposes [illegible] the timber cut and man[illegible] in Virginia was valued at $6,172,312.

[illegible] Virginia, like those of all maritime or [illegible] states, are of two kinds: 1. *Oceanic* [illegible] more or less saltish and in the [illegible] generally superficial and [illegible] and the tributaries and [illegible] derived, but sometimes sub[illegible]

[illegible] peculiarly rich in waters of both these [illegible] portion of [illegible] will show. *Tide-water* [illegible] a front of nearly 120 miles [illegible] Sea and the great ocean [illegible] beyond it over a marine [illegible] chartered jurisdiction [illegible] Chesapeake Bay, the

only Mediterranean of the United States, which spreads itself and a half dozen of its great and hundreds of its smaller tidal arms through more than a fourth of the territory of the State. *Its fresh-water rivers* are gathered by a score or more of important rivers and their branches, flowing in all directions and draining portions of five great catchment basins, from tributaries and springs well-nigh innumerable.

Its tidal ways are abundantly developed and admirably adapted to navigation, and many of them are broad, deep, and land-locked and land-protected estuaries in which the navies of the world might take refuge.

Its fresh-water rivers and their tributaries, unrivaled in number, meandering through every portion of the State, are generally well supplied with water during every season of the year, as they should be in this region of copious perennial rains and where the geological conditions are mostly favorable for retaining the precipitation on or near the surface. Nearly all of these have a rapid descent from the successive plain and mountain terraces of their sources, so that they not only water the land and

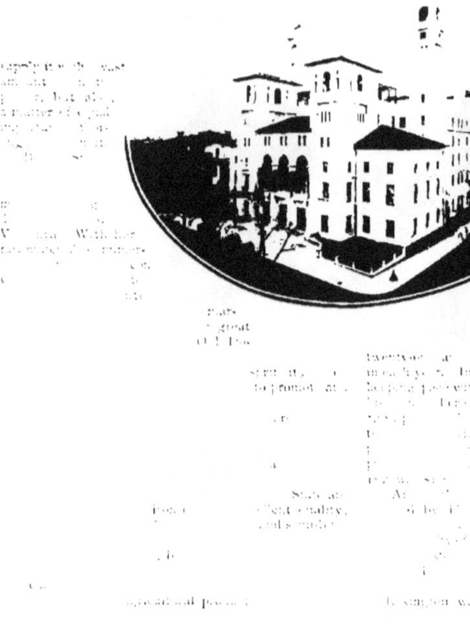

the Miller Manual Labor School at Crozet; the Hampton-Sidney College at Hampton-Sidney; the Roanoke College at Salem; the Emory and Henry College at Emory; the Episcopal Theological Seminary; the Union Theological Seminary, Hampden-Sidney; the Martha Washington College and Conservatory of Music; the Southern Female College; the Randolph-Macon College for young ladies at Lynchburg; the Polytechnic Institute at Blacksburg, and several female seminaries at Staunton.

It will thus be seen that the most ample provision is made by Virginia for education, from the public school to the university, and for all branches—agricultural, mechanical, scientific, literary, and professional. In Virginia more than in any other State the old English method of private boarding schools still flourishes.

The cities of Virginia have ever been to the old mother State among her most precious jewels. The mother of the Gracchi was not more proud of her sons than is the Old Dominion of these daughters. They were centers of patriotic activity in earlier days, just as in this later time they have become busy centers of commerce and of manufacturing.

Richmond, of course, by reason of her history, her population, her splendid progress, is the first city and capital of the State and one of the first cities in the South. The glamour of her past does not [illegible] her to the glory of the future. She is pulsating with the broad upward movement that is everywhere advancing

THE GREAT SEAPORT TERMINALS OF THE SOUTHERN RAILWAY AT PINNER'S POINT, NORFOLK, VA.

The magnificent Hotel Jefferson at Richmond, erected at a cost exceeding $1,000,000 by the late Major Lewis Ginter, is one of America's most palatial hotels. It has become a most popular resort for tourists and travelers, who find it perfect in all of its appointments.

Across the river from Richmond is the city of Manchester, a brisk manufacturing center of 10,000 inhabitants. It has superb water power which is largely utilized. Here too are located repair shops of the Southern Railway.

Norfolk is the largest port on the Atlantic south of Philadelphia. In her splendid harbor at the head of Hampton Roads, one of the finest in the world, are seen the flags of every maritime nation. Her commerce extends to every sea. She is the great water gateway of the South, through which the products of this mighty empire seek a market. Like Venice in the middle ages in all of her power, Norfolk is a modern "bride of the sea." The value of her exports for 1897 was [illegible]. In [illegible] her exports amounted in value to [illegible]. There are nineteen lines of steamships engaged in the coast and foreign trade with Norfolk as

in the South [illegible] with [illegible] growth [illegible] population [illegible] increase [illegible] per cent [illegible]. Her present population is estimated at [illegible]

[illegible]

[illegible] a terminal. One of these is the New Bay Line, operating [illegible] steamers between Baltimore, Norfolk and Old Point Comfort.

[illegible] supported by [illegible]

Across the harbor from Norfolk is Pinner's Point, a [illegible] terminal of the Southern Railway. Here the railway [illegible] three wharves, one 106 by 800, one 272 by [illegible], and [illegible] by 800 feet, giving a total wharf-[illegible] of [illegible] feet. Of freight sheds there are four, one [illegible] by [illegible], one double shed 200 by 400, and a fourth

252 by 700 feet, giving a total capacity of shed room of thirteen acres. Along these docks are railroad tracks to carry freight to the steamship's side. Five dock slips are already in existence, 200 by [illegible] feet each. There is a depth of water of twenty-seven feet at the [illegible] wharves, which will accommodate boats of the largest class. No railway in America [illegible] better terminals, and there are but few such in the world. These warehouses, piers [illegible] docks [illegible] almost [illegible] [illegible]. With its thousand laborers, its own water works and electric light plant, its own modern fire department and alarm system, with its miles upon miles of sidings, its powerful compress, [illegible] system [illegible] of [illegible] [illegible], but [illegible] whatever the volume of work and seeming hurry it is a most fascinating city either to the layman or transportation expert who finds himself within its magic midst. It is a striking example of what energy, when coupled with capital, can do, for all of this hive of industry but little more than a year ago was nothing more than a swamp, and where busy engines puff to-day, fifteen months ago tall reeds nodded and bowed to the wind. Through Pinner's Point (for this is the local name for the Southern terminals) [illegible] great volume of traffic between the North and the South, the West and the Southwest. From here sail steamers not only to the coast cities of the Union but to the ports of almost every nation. Nearly one half of Norfolk's cotton [illegible] (and Norfolk ranks [illegible] among the Southern cotton ports) passes over its piers. Merchandise is handled here the value of which would stagger the mind.

Norfolk is a growing grain [illegible], especially in corn. In [illegible] the receipts of corn amounted to [illegible] bushels. In 1895 there was an [illegible] [illegible]

[illegible] centres of the country. The receipts of wheat and oats in 1896 [illegible] bushels, a good increase over previous years. [illegible] market in the world.

In [illegible] business, Norfolk has an immense coasting trade [illegible] hundred vessels leave her harbor every month with [illegible] fruits and vegetables, more

in 1776. The tombstones in the surrounding churchyard bear epitaphs dating back as early [illegible] 1673, and mark the [illegible]sting place of many of Virginia's earliest and most honored sons.

Buoyed by a past of [illegible]oried greatness, Norfolk is pressing forward to the coming century with a dauntless faith in yet larger greatness in commerce, in industry, in every high work of progressive civilization.

Nearby Norfolk are two of the most famous resorts in America, Old Point Comfort and Virginia Beach. The former is upon the historic waters of Hampton Roads, which is formed by the confluence of Chesapeake Bay and the James River. There are two hotels here, the Chamberlin, said to be the finest hotel on the Atlantic Coast, and the always popular Hygeia. The hotels adjoin Fort Monroe, one of the largest of the Government's military posts, and overlook the beautiful sheet of water which was the

[illegible] during the [illegible]

[illegible]

[illegible] naval duel between the Monitor and Merrimac, and which is now the winter station of the White Squadron. The peculiarly delightful climate, added to the brilliant social life, has made Old Point a most popular resort in winter [illegible] Northern people and in summer for visitors from the South.

Virginia Beach, at which there is a [illegible] hotel, the Princess Anne, is [illegible] miles east from Norfolk and directly upon the ocean. This beautiful [illegible]ort is a favorite rendezvous for people from Southern cities [illegible], and the hotel is always [illegible] with guests during the winter [illegible] New York and the North.

A[illegible] of Virginia W[illegible] Norfolk at

NEW QUADRANGLE OF THE UNIVERSITY OF VIRGINIA, FROM ARCHITECT'S DRAWING.

shipping business in cotton, flour and lumber is done with North Atlantic ports, Europe, and South America. The water is so deep that the largest vessels move about easily. West Point is but twenty miles from Richmond. Its situation is excellent for various kinds of manufacturing and for general business. King William County, in which it is situated, is mainly agricultural, having some of the richest farming lands in Virginia. All grains, tobacco, and vegetables flourish. The territory tributary to West Point is especially adapted to truck farming, and the waters abound in oysters and fish, which form a considerable portion of the town's industry, immense quantities being shipped to Northern markets daily. The climate is excellent, the average temperature being 57 degrees and the rainfall about 42 inches.

SOUTHSIDE FEMALE INSTITUTE, BURKEVILLE, VA.

Turning from the Tidewater region to beautiful Piedmont and going south on the main line of the Southern Railway one reaches Charlottesville, the seat of the University of Virginia, a charming little city, whose academic atmosphere is tinged with the mellow light of a glorious past. At nearby Monticello lived Thomas Jefferson, the founder of the university and the author of the Declaration of Independence. Like Mount Vernon, it is a Mecca for every patriot, and the present owner is always glad to give visitors the privilege of seeing the historic old homestead. Besides the University of Virginia, which annually expends in the community $350,000, there are located in Charlottesville the Piedmont Female Institute, Albemarle Female Institute, Charlottesville Seminary, the Miller Manual Labor School, Pantop's Academy and Jones's Classical School. In all the South there is no city with more advanced educational institutions. While these may be termed the city's chief industry, Charlottesville also has the distinction of having the largest woolen mill in the South and of producing wines from the clustering vineyards of Piedmont that have won a world-wide and enduring fame. There are, too, factories of various kinds that hum with prosperous industry. It is a city wherein knowledge is the handmaid of industry, both making for the best things in mind and in matter.

A LEAF TOBACCO AUCTION SALE

Continuing South from Charlottesville through the charming Piedmont region, which bespeaks prosperity and wealth on every hand, one comes to the fine old city of Lynchburg, situated on the banks of the James, approximately in the center of the State. The country tributary is noted for its fertile soil and uniform climate. The city has [illegible] inhabitants, and is growing steadily in commerce and manufactures, as well as in population. It is a busy jobbing center, having a large number of wholesale houses, which do an annual business of [illegible]. Lynchburg has long been famous for its tobacco trade, the total sales of leaf tobacco being annually about [illegible] pounds. The banking capital of the city is [illegible], and over [illegible] firms are engaged in business. Its superior railroad facilities make it a natural assembling point for the products of mine, forest and field, and offer cheap transportation of the manufactured product to market. These advantages are attracting increasing attention to Lynchburg as a center for profitable manufacturing. It is a busy, thriving city, pushing ahead on all lines of enterprise and industry.

The growth of the city has not been in the nature of a boom, but upon the basis of steady increase from energy and enterprise. In all matters that affect the city's welfare, Lynchburg is fully abreast of the times. It has its streets paved, wherever practicable, with Belgian block. The city is lighted by electricity, and, notwithstanding the steepness of the hills, an electric railway passes around the entire city.

The city is connected with the town of Madison by a [illegible] iron bridge across the James. The city is also connected with the suburb Rivermont by a splendid iron bridge over Blackwater Creek, 1,2[illegible] feet in length, [illegible] feet wide and 132 feet high. It carries a [illegible] electric railway, two roadways 20 feet wide, and [illegible] walkingway on each side. This bridge connects with the great avenue, [illegible] feet wide, upon which is [illegible] Randolph-Macon Woman's College, designed to give young women all [illegible] advantages that Randolph Macon gives the young men.

[illegible] of Lynchburg, in which the city takes the highest pride. [illegible] The city in every way combines the [illegible] of a prosperous commercial center.

[illegible] has nowhere been embodied more fully than in [illegible] Dan River into the enterprising city of Danville. [illegible], the banks of which are now occupied by [illegible] of trade and industry now has a population of [illegible] for loose leaf tobacco in the world, and, with

possibly one exception, the largest brightleaf tobacco market in America [illegible] sales of [illegible] pounds [illegible] year ending [illegible] The [illegible] was [illegible] the aggregate [illegible] not include at least [illegible] pounds purchased by Danville leaf dealers elsewhere. Here [illegible] located some of the largest [illegible] in the South, with [illegible] aggregating [illegible] and with an annual [illegible] productive [illegible]

[illegible] industries give employment to [illegible] and hundreds more find remunerative labor in the planing mills [illegible] and in the city's large jobbing trade, which is [illegible] increasing. [illegible] The [illegible] that could possibly be [illegible]

[illegible] with North Danville by a splendid [illegible] bridge [illegible] in their interests and business relations. The streets are well lighted by [illegible]. The water, gas and electric light plants all belong to the city, which does not attempt to make money by the enterprises, but to furnish light and water at prices approximating cost. There are many handsome buildings, a fine new city hall, United States public building, a market-house, fine public school buildings, two bridges [illegible] the river, and a large suburban park now being attractively improved. There are also electric railways, [illegible] exchange, free delivery of mails, beautiful theatre, and good hotels. The total value of church property [illegible] and it is a remarkable fact that fifty per cent of the population are church members.

The school system of the city is most excellent, and [illegible] colleges are doing much to advance the [illegible] young women. The Danville Military Institute has long maintained a [illegible] institutions of [illegible]

Danville [illegible] attractive [illegible] town, and its people are [illegible] Danville has [illegible] working works. Her [illegible] and [illegible] steadily growing.

With Danville [illegible] going on into North Carolina. The brief [illegible] of Virginia's resources [illegible] that [illegible] will at least serve to show that the State is awake to her opportunities. [illegible] behind her are three centuries of splendid history, before her is a dawning twentieth century, [illegible] Virginia [illegible]

THE SWANNANOA RIVER NEAR ASHEVILLE—LAND OF THE SKY

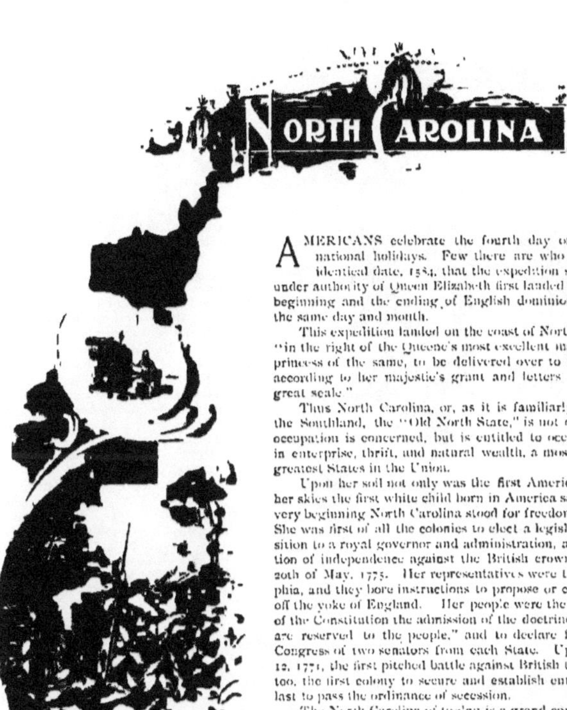

NORTH CAROLINA

AMERICANS celebrate the fourth day of July as one of their great national holidays. Few there are who recall that it was upon this identical date, 1584, that the expedition sent out by Sir Walter Raleigh under authority of Queen Elizabeth first landed upon American soil. Thus the beginning and the ending of English dominion in this country occurred on the same day and month.

This expedition landed on the coast of North Carolina and took possession "in the right of the Queene's most excellent majestie, as rightful queene and princess of the same, to be delivered over to the use of Sir Walter Raleigh according to her majestie's grant and letters patent, under her highnesses great seale."

Thus North Carolina, or, as it is familiarly known among its sisters in the Southland, the "Old North State," is not only the oldest so far as white occupation is concerned, but is entitled to occupy, by right of her prowess in enterprise, thrift, and natural wealth, a most prominent place among the greatest States in the Union.

Upon her soil not only was the first American colony founded, but under her skies the first white child born in America saw the light of day. From the very beginning North Carolina stood for freedom and the rights of the people. She was first of all the colonies to elect a legislature by popular vote in opposition to a royal governor and administration, and the first to make a declaration of independence against the British crown, that of Mecklenberg on the 20th of May, 1775. Her representatives were the first of all sent to Philadelphia, and they bore instructions to propose or concur in the movement to cast off the yoke of England. Her people were the first to demand in the framing of the Constitution the admission of the doctrine that "all powers not granted are reserved to the people," and to declare for an equal representation in Congress of two senators from each State. Upon her soil at Alamanca, May 12, 1771, the first pitched battle against British tyranny was fought. She was, too, the first colony to secure and establish entire religious freedom, and the last to pass the ordinance of secession.

The North Carolina of today is a grand commonwealth of 2,500,000 population, rich in all that goes to make for human progress, and possessing a wealth of minerals, timber and fertile lands which are being turned rapidly by the energy and enterprise of her citizens into money riches. She has reached a property valuation as listed for taxation of $230,8[illegible]1,131, of which $8,1[illegible]0,074 is to the credit of her colored citizens. She has 3,577 miles of railroad, having an assessed valuation of $24,555,754, within her borders. There are in successful operation 1[illegible]2 cotton mills, 17 woolen mills, with a half dozen more building, 220 tobacco factories, and over [illegible] miscellaneous manufacturing establishments.

In all lines of human progress, North Carolina's development has been wonderful. Her State University, located at Chapel Hill, was the first State university to be established, and holds high rank among the best educational institutions of the country. She has a most comprehensive public school system, for the support of which the State appropriates nearly a million dollars annually. She maintains normal schools for colored pupils at Salisbury, Fayetteville, Goldsboro, Plymouth, Elizabeth City and Winston-Salem, and has

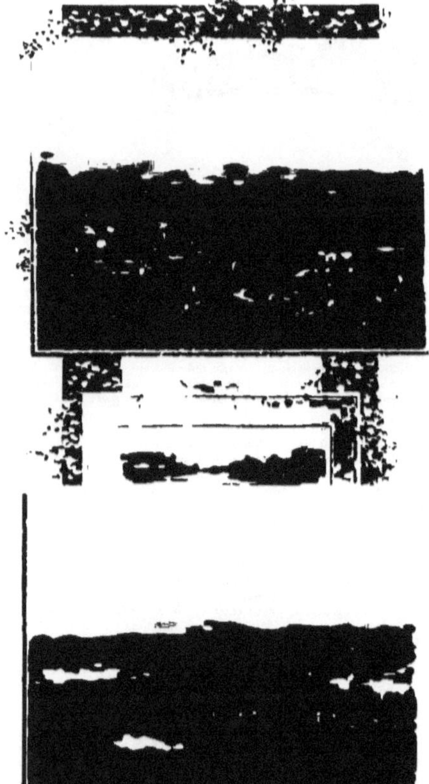

at Asheville, the St. Mary's College for girls at Raleigh, and the Salem Female Academy at Winston-Salem. It is not strange that a State in which but one-fifth of one per cent. of the population are of foreign birth, and ninety-five per cent. are of State nativity, should be alert in the education of its young.

Geographically North Carolina is an empire in itself. Its total length is 5[illegible] miles, and it has an area of 52,250 square miles, of which [illegible] per cent. is forest. It would hold ten States the size of Connecticut and six as large as Massachusetts. It has a greater diversity of climate than any State except California, and could approximate more closely the maintenance of its inhabitants, independent of outside markets or products, than any territory of equal size in the world.

There are in North Carolina three great physiographic divisions or terraces, the Coastal, Piedmont and Mountain. The White Mountains are dwarfed in comparison with the sublime heights in the western or mountain region of the State, where forty-three distinct peaks attain a higher altitude than Mount Washington, and over eighty approximate it in height, the mean altitude being greater than any section east of Colorado. The middle portion, known as the Piedmont plateau, is a wide-stretching, undulating region of fertile farm lands unsurpassed anywhere for agrarian purposes, while the eastern or coastal plain is rich in waterways and in a soil productive to the highest degree.

Reference to the mean parallels of latitude will show that North Carolina is situated nearly midway [illegible] the Union; and inasmuch as the Union lies entirely [illegible] temperate zone, it follows that North Carolina is situated upon the central belt of that zone. This position gives to the State climatical conditions and productive capacities not excelled by any in the world. As a [illegible] writer has put it, "the Old North State is the [illegible] of Summer and Winter." On the west the [illegible] mountain chains interpose their mighty barrier between the bleak winds of the northwest and the general surface of the State. On the east the coast is swept by the Gulf Stream, the meliorating effect of which is [illegible]. From this position and these causes the [illegible], which is more or less the life of all vegetation, ranges within moderate limits from season to season. Including all the sections heretofore named, the range of climate in North Carolina is the same as [illegible] the Gulf of Mexico to the Gulf of St. Lawrence. [illegible] of this fact is seen in the wide range of [illegible] products freely growing within [illegible] the palmetto and magnolia grandiflora [illegible] hemlock and balsam fir, and from [illegible] to Canadian oats and buckwheat. [illegible] every product found between the Great [illegible]. With an average mean temperature [illegible] freedom from torrid heat or the [illegible]. The skies rival in their azurine [illegible] and there is a vitality and tone in [illegible] an instant impression on [illegible]

[illegible] ation of advantages gives to the [illegible] world [illegible] climate [illegible] the purity of the translucent, [illegible] which nearly all [illegible] flow and

There are in North Carolina 3,[illegible] miles of rivers, 1,[illegible] miles of which are navigable. The seven principal streams, the Roanoke, Tar, Cape Fear, Neuse, Yadkin, Catawba and French Broad, have an average fall of ten feet to the mile, and furnish an estimated aggregate horse power of over [illegible].

In agriculture the State takes high rank, and offers an inviting field for settlement and enterprise. No other region in the United States presents so many attractions in the farming line to the man of large or moderate means as the Piedmont or mountain region along the line of the Southern Railway. This is true because of the wonderfully fine climate, the picturesqueness of the scenery, the magnificent sites on mountain sides, where views of miles of lofty mountains may be had, and the more important fact that from the richness of the soil, the great variety of grasses, the abundance of pure water, the peculiar purity and richness of the air, there is the best of opportunity for cotton, tobacco, fruit, grain or stock farming on large or small scale.

In cotton culture North Carolina takes a prominent place among her sister States of the South. There are but eight counties of her ninety-six in which it is not grown, and the area devoted to it is considerably over a million acres, the soil of the State being particularly well adapted to its growth. In the manufacture of cotton North Carolina has [illegible] There are in the State today [illegible] [illegible] several [illegible] the State [illegible]

[illegible] within the [illegible] of the State, [illegible] to this [illegible] which [illegible] North Carolina [illegible] of any of the States. [illegible] to the mineral side of North Carolina's natural [illegible] [illegible] Copper, silver, iron, [illegible]

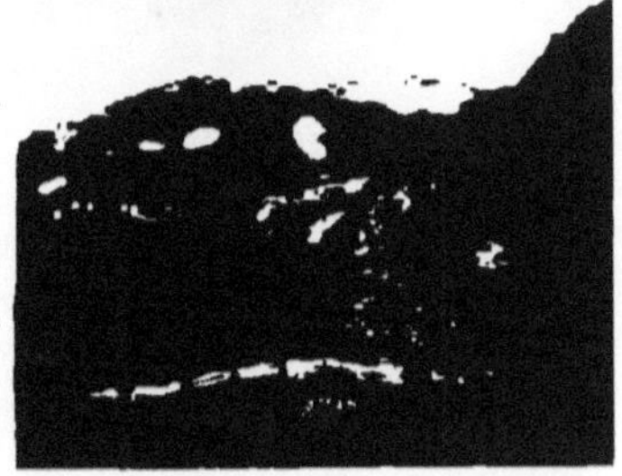

the timber y[illegible] s a revenue to her hardy [illegible] of over $[illegible] annually. Originally [illegible] magnificent forests [illegible] density [illegible] savannahs [illegible] western boundaries. Upon the [illegible] on the Piedmont [illegible] hard woods with [illegible] are found [illegible] the United States [illegible] deciduous [illegible]

Mr. G[illegible] Vanderbilt, at his magnificent estate near Ashe[illegible] has established a forest department under [illegible] and competent management, and is carrying [illegible] not only in [illegible] the timber in his [illegible] acres, but in generally [illegible] of [illegible] the State.

Th[illegible] North Car[illegible] nearly 6,000 [illegible] that of [illegible] State [illegible] territory [illegible] the w[illegible] For [illegible] years the

estimate [illegible] With [illegible]

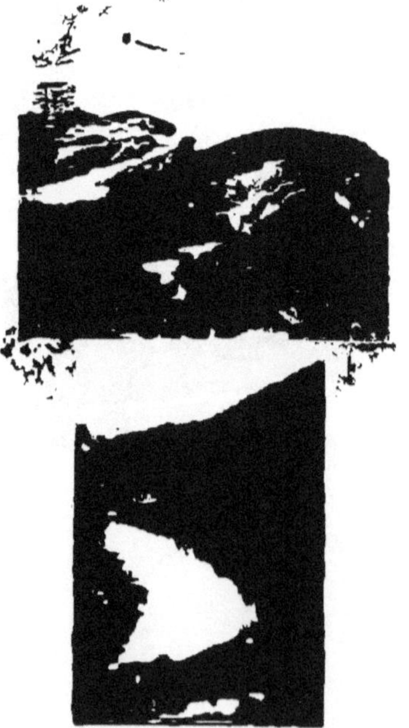

[illegible]

State has been the source of the national supply of crude vegetable drugs. More than seven hundred distinct species of important medical plants grow wild in the State, and furnish an industry the volume and importance of which is appreciated by few outside of the medical and pharmaceutical professions.

North Carolina is not alone great in her industrial and material wealth; she is superb, majestic, sublime in all of those qualities which awaken in man the heart-throbbings of enthusiasm over the stupendous works of the Almighty, as portrayed in towering mountains and deep-shadowed gorges.

Europe may have her Switzerland, the West [illegible] Colorado, the Pacific coast may glory in her Sierra Nevada and British Columbia in her Cascade range [illegible] nowhere on the face of the earth is there a region more picturesquely, more charmingly beautiful than the mountain country of western North Carolina, poetically known as "The Land of the Sky." It is true there are mountains of greater elevation in each of the localities named, but the greatest canvases in the gallery of art are not the choicest gems, nor is the beauty of nature to be measured on opposite lines. Where the mountain ranges of the West are rugged, barren and [illegible], th[illegible] in western North Carolina are robed in deep-hued forests to their [illegible] summits. Where the great [illegible] of the Sierra Nevada frown, th[illegible] "The Land of the Sky" smile [illegible] with [illegible] of [illegible]

Where the valleys of the one are rocky and impassable gorges, in the other they are fern-carpeted forest labyrinths, through which crystal streams tumble merrily along over moss-grown rocks in their race to the open.

Picture in your mind a region where range after range of heavily forested mountains parallel each other like waves of the sea, where interlacing valleys are rich with verdure and flowers, and where silv[illegible] streams murmur unceasingly. Imagine an air so [illegible] and pure that breathing itself seems a new-found joy, then throw over all a canopy of bluest of Italian blue, and you have "The Land of the Sky."

"Land of forest-clad mountains, of fairy-like streams,
Of low, pleasant valleys where the bright sunlight [illegible]
Athwart [illegible] over the hills,
[illegible] the fragrance of pines and the murmur of rills.

"A land [illegible] bright sunsets, whose glories extend
From [illegible]
The [illegible] the rainbow with clouds [illegible]
Right well art thou christened! The Land of the Sky.

"A land of pure water, as pure as the air;
A [illegible] for the fair;
[illegible]
[illegible] from [illegible]

"As far from [illegible] North as from the [illegible]
[illegible]
[illegible]
[illegible] The land of the sky

[illegible]

The mountains that [illegible]
[illegible]
[illegible] of the ages long past
[illegible] valleys below
Where [illegible] crystal streams [illegible]

"The pure [illegible] the [illegible]
[illegible]
[illegible]
[illegible] Land of the Sky."

This rugged mountain region embraces the extreme western portion of North Carolina and the eastern edge of Tennessee. Within these confines are several districts, alike in their general features, but each having distinct charms and advantages peculiarly its own. The one [illegible] generally visited has Asheville for its tourist center. None the less beautiful, however, is that country in and about Blowing Rock and Grandfather Mountain, of which Lenoir is the *[illegible]*. Southeast of Asheville is the Flat Rock and [illegible] region which attracts

Carolina was erected at Greensboro, and now there are several mills here, [illegible] in the State. There are extensive [illegible] factories, planing mills, and many other prosperous establishments among Greensboro's enterprises. The notable buildings in the city include the United States Government building and county courthouse, the latter one of the finest [illegible] in the State. [illegible] the city is especially rich. It is the seat of the Greensboro Female College, with [illegible] students; the State Agricultural and Mechanical College for colored youth, with [illegible] students; the State Normal and Industrial School for young women, with [illegible] students; Bennett College for colored youth, with [illegible] in the vicinity [illegible] Guilford College, a well-known [illegible] institution, with [illegible] students. Guilford County [illegible] originally settled by the Scotch-Irish, and the [illegible] properties of these early pioneers [illegible] community.

Fifteen miles south of Greensboro, on the main line of the Southern Railway, is High Point, where the Sportsmen's Association holds the Eastern Field Trials each year. Its population [illegible] in eight years. It is [illegible] center, and has a [illegible] industries, six churches [illegible].

The next place of importance [illegible] of High Point is Salisbury, a town of [illegible]. It has several manufacturing [illegible] mill. The [illegible] the most prosperous [illegible]

sections in North Carolina, and has an assessed valuation of $2,07[illegible]. One of the national soldiers' cemeteries is located at Salisbury, and it is also the seat of one of the state normal schools and Livingstone College. Because of the salubrity of its climate a modern sanitarium is projected.

One mile and a half north of the city on the main line is the new town of Spencer, named in honor of Mr. Samuel Spencer, the President of the Southern Railway. Here the road has erected extensive shops employing a large number of men, and established division headquarters. Naturally this young town, which is but a year old, has had a wonderful growth, and all indications point to its becoming a prominent city within the next few years. The Southern Railway is extending its shops rapidly, and about these there will grow up a prosperous community. Rowan County, in which Spencer is located, is one of the finest in the South, and with a rich agricultural backing and a substantial industrial foundation such as the Southern Railway shops will give it, the future of Spencer is particularly bright.

Twenty-three miles south of Salisbury is Concord, one of the progressive smaller cities of the State. It has [illegible] mills and is growing rapidly. Its population [illegible], alert and enterprising.

Charlotte, the midway city on the Southern's main line between New York and New Orleans, is [illegible]

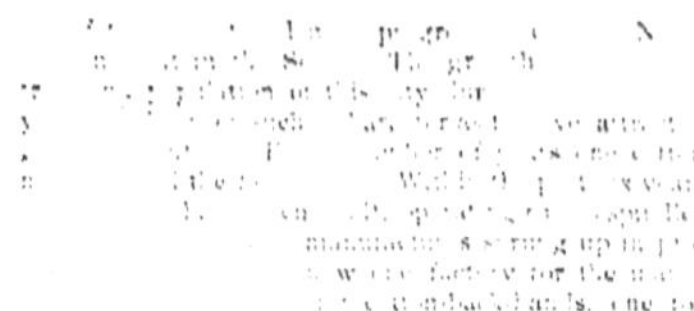

and wood-working shops, and a fairly good supply of all the smaller industries of a thriving town. There are employed in the various factories of the city 3,500 people, who draw in wages $1,[illegible] per year. The population has kept pace with the growth of the manufacturing enterprises. In 1880 it was [illegible]. In [illegible] it had increased to [illegible], and a city directory issued in August, [illegible], showed a population, including the suburbs, of [illegible].

The streets of the city are electric lighted, and the sewerage system, which extends to all parts of the town, is of the most approved type, as it is aided by the topography, the ground sloping down to the swift-running

streams which bound the city on its east and west sides. The city's garbage is disposed of by cremation.

Charlotte is particularly fortunate in its handsome buildings, both public and private. The government's post-office and court house building cost $[illegible]; the [illegible] and handsome county court house each cost an [illegible], the latter being the finest county building in the State. The residences erected in recent years are of the finest type of architecture and attract the attention of all visitors to the city. The city has fine sub[illegible] banks, with an aggregate capital of $1,250,000, and [illegible] valuation is $[illegible].

[illegible] center of a rich gold-mining section [illegible] United States Assay Office at this place daily [illegible] from the mines of North

Carolina, South Carolina and Georgia. It is the seat of Elizabeth College for women, Biddle University for colored students, and the Presbyterian College for women, and in addition to its excellent public schools has several private educational institutions.

There is to be erected in the near future in front of the court house a monument to the signers of the Mecklenburg Declaration of Independence, as this was the spot where, on the 20th of May, 1775, the convention called for the purpose first formally renounced allegiance to England. This Declaration antedated the one at Philadelphia by more than a year.

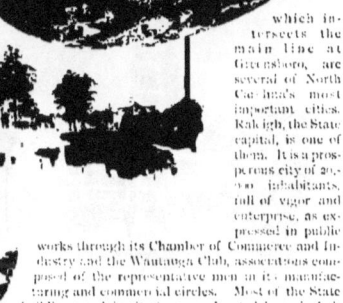

Between Charlotte and the South Carolina line is the prosperous town of Gastonia, a place of 3,000 inhabitants, full of energy and thrifty enterprise. It has had a strong and vigorous growth during the past few years. To-day it prides itself on its four cotton factories and other industrial establishments. It has eight churches and good graded schools.

Upon the line of the Southern Railway from Norfolk, which intersects the main line at Greensboro, are several of North Carolina's most important cities. Raleigh, the State capital, is one of them. It is a prosperous city of 20,000 inhabitants, full of vigor and enterprise, as expressed in public works through its Chamber of Commerce and Industry and the Wautauga Club, associations composed of the representative men in its manufacturing and commercial circles. Most of the State buildings and institutions are located here, including, beside the classic capitol building, a superb State Museum and the State Library. The city has nearly fifty miles of broad, well-paved and well-shaded streets, and a fine water and sewerage system. Raleigh has just reason to be proud of the fact that the issue of her street improvement five per cent. bonds was recently sold at one hundred and ten, the highest price ever realized from the sale of Southern municipal securities. The manufacturing industries of the city embrace a large hosiery, yarn and gingham mill, phosphate works, a cotton-seed oil mill and tobacco factory.

Raleigh is one of the leading educational centers of the State, having three colleges for young ladies: Peace Institute, St. Mary's (now in its fifty-sixth year), and the Baptist University; the Agricultural and Mechanical College; the State Institute for the education of the deaf, dumb and blind, a male academy, and a fine system of modern public schools.

For the higher education of the colored people of the State there are Shaw University and St. Augustine Normal College.

North Carolina ... for years one of the most ... agricultural ... any State in ...

Durham's representative citizens, Mr. Washington Duke and Mr. Julian S. Carr, and the splendid graded schools, with about [illegible] boys and girls in daily attendance.

Between Raleigh and Durham [illegible] the junction known as University, from which a branch a few miles in length runs to Chapel Hill, the location of the University of North Carolina, the leading educational institution of the State, chartered [illegible] in 17[illegible]. It is the oldest university in the South and the oldest State university in the Union. Its roll of alumni includes seven thousand [illegible]. Many of them are of national repute, and it may be doubted whether so large a percentage of the alumni of any other American college have achieved [illegible] in public life. The university embraces the [illegible] school, the medical school, the school [illegible], and the summer school. The college [illegible] departments, offering [illegible] courses of instruction [illegible] both graduate and undergraduate [illegible]. The university includes [illegible] teachers, who [illegible] the training of 22 American and European [illegible]. The student roll numbers [illegible], the summer [illegible]; total [illegible]. It possesses property worth about [illegible] museums, laboratories, and student rooms. The library contains 40,000 volumes and pamphlets. The gymnasium is the largest in the South. The income is about $50,000 a year. The university is administered with great economy; total expense of an education for one year need not exceed $200. The president of the university is Edwin A. Alderman, D. C. L.

Goldsboro is the terminus of the branch of the Southern Railway starting at Selma, twenty-seven miles east of Raleigh. It is a progressive, active place, the center of a prosperous agricultural country. Its population is about seven thousand, and is steadily on the increase. It is the county seat of Wayne county, and is a growing manufacturing center. It has one of the largest lumber plants in the South, an extensive furniture factory, a cotton mill, recently equipped with the very latest improved machinery for both spinning and weaving; one of the most noted fertilizer works in the country, a cotton-seed oil mill, a rice mill, and numerous other manufacturing establishments. The city is absolutely free from debt, and has a good fire department, water works, electric lights, paved streets, churches of every denomination, and excellent public graded schools for both white and colored, with an attendance of [illegible].

GLIMPSES OF GASTONIA, N. C.

Oxford, upon the branch running north from Durham, is the center of a large tobacco and cotton region, and is the seat of a Baptist female seminary. The town has several manufacturing enterprises, and enjoys considerable local trade.

Twenty-nine miles west of Greensboro is the enterprising and prosperous city of Winston-Salem. Its present population is 20,000, an increase from 4,194 in 1880. It has a larger number of plug-tobacco factories than any city in the world, and the purchases of the leaf reach the aggregate of about $2,000,000 annually. In a single year the manufacturers of the city have paid the United States Government for revenue stamps almost $1,000,000, and it is claimed that more money is disbursed here in wages annually than in any city in the South equal in population. One of the city's latest enterprises is the transmission of electric power for her extensive manufacturing establishments from the Yadkin River. There are fine streets and all urban improvements, a superb municipal building, a model Y. M. C. A. building, a chamber of commerce, handsome churches and most excellent public schools. The Salem Female Academy, located here, is not only the pride of the city but one of the famous institutions of learning of the South. It was established by the Moravians in 1802, and at least 10,000 alumni claim it as their alma mater. Two of its graduates have graced the White House at Washington, Mrs. President Polk and Mrs. Patterson, daughter of President Jackson. In addition to the academy there are located here the Slater Industrial Academy and Normal School and the Davis Military Academy.

Upon the line of the Southern Railway running west toward Asheville from Salisbury, and between the two places, are several important towns and some sublime mountain scenery. Statesville, a place of 3,500 inhabitants, has a large cotton mill of 6,000 spindles running day and night, several large tobacco factories, two steam flour mills and a number of cooperage establishments. In addition to its excellent public schools, there is located here a female college occupying a handsome edifice of its own.

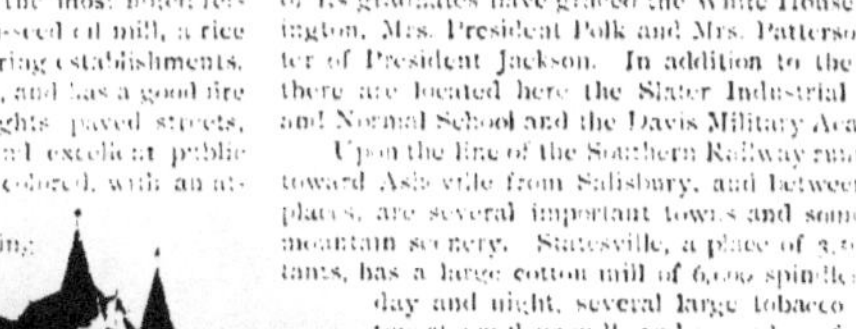

STATE INSTITUTIONS FOR DEAF, DUMB AND BLIND, RALEIGH, N. C.

South of Statesville, on the branch of the Southern Railway connecting it with Charlotte, is Davidson College, one of the best known of North Carolina's educational institutions. It is a Presbyterian institution, and has had for many years a most prosperous career. It is well endowed and has a large and progressive faculty, and many students attend not only from North Carolina but from other States.

Beyond Statesville is Hickory, a popular place for sportsmen and an enterprising town. It has an excellent hotel, the Hickory Inn, which enjoys a considerable tourist patronage. It is also the location of St. Paul's Seminary, a Lutheran theological institution. The surrounding country is attractive beyond description and one of the best [illegible] regions in the State. It is at Hickory that the tourist leaves the Southern for Lenoir and the wonderfully [illegible] region about Blowing Rock and Grandfather's Mountain, to which reference is made elsewhere in this chapter.

M[illegible] some [illegible] miles beyond Hickory is a pretty [illegible] town, beautifully located [illegible]ng the lower mountains, and it is here where the railroad fairly begins the ascent of the great mountain range.

From here to Asheville, and thence to the Hot Springs, the traveler does not pass over a mile of uninteresting territory. As the train begins its [illegible] ascent of the mountains, which seem to be piled up in impassable masses, the scenery [illegible] finally, as the summit is approached [illegible]. The [illegible] of iron and brass [illegible] mountain fastness [illegible] to the very edge of the [illegible] and in a moment [illegible] a dark, [illegible] cleft [illegible] beneath [illegible] a stream [illegible] its way to the valley. [illegible] rapidity. Up and [illegible] the summit [illegible] and dwarf the nearer [illegible] different places [illegible] at your feet. One brief and [illegible] into the Swannanoa [illegible] mountains. You enter it [illegible] the Mississippi and the [illegible] Crystal Spring, in its center, as [illegible]

legislature has granted a charter with ample powers and generous privileges. The enterprise is a community in the sense that all profits from the sale of lots will be used for the benefit of the entire community and for the purposes for which taxes are usually laid. By the charter the sale of intoxicating liquor is forever prohibited. Plans are being made for a large and important educational institution, one hundred acres of land having been set aside for this purpose, and the resort will also be a center for annual gatherings of prominent and earnest Christians at work for the study of problems relating to the welfare of humanity, and ways and means for advancing the interests of Christianity through the various denominations.

The descent from Black Mountain to the level of the Asheville plateau is gradual and the passing scenery beautiful. Just where the railroad meets the lovely Swannanoa River is the handsome station of Biltmore, at the very corner of the vast estate of Mr. George Vanderbilt.

Two miles beyond is Asheville, which is the tourist as well as the commercial center of this region. It has been called the "Janus of resorts," for, like that two-faced divinity of the ancient Romans, it has two fronts. Upon one it wears a welcome for the winter guests from the North, and upon the other a smiling greeting to the thousands who come here each summer from the Southern cities to enjoy the cool, bracing air of the mountains.

Asheville has a greater elevation than any city east of Denver, being 2,300 feet above sea level. It occupies an ideal site just at the merging of the ever-beautiful Swannanoa (nymph of beauty) River with the historic French Broad. The mountains have drawn away, leaving as fair a valley or plateau as human eye ever gazed upon. But raise your eyes in any direction above the immediate surroundings of undulating hills which have been left by the erosion of the rivers, and they will rest upon the circling ranges of towering mountains, which give a glorious setting to the picture. The city of Asheville has had a vigorous growth. It has an active air of commercial life, and upon every turn there are indisputable evidences of thrift and prosperity.

Considered from a business and manufacturing point of view, the place occupies an enviable position among North Carolina's cities, and as a tourist center its fame is world wide.

Th[illegible] and electric cars run through the [illegible]

schools, maintained by private subscription, and the Asheville Farm School, occupying [illegible] acres, which has over one hundred students who are taught agricultural work on approved and practical scientific lines.

pa[illegible] far[illegible] attractive suburbs. T[illegible] city [illegible] a country club a good [illegible] and a public library. The ch[illegible] buildings are modern and well [illegible] educational cen[illegible] well known institutions of [illegible] public schools. The [illegible]

Asheville has a permanent population of about [illegible], and there are always a large number of visitors, estimated to average several thousand.

The business portion of the city centers about the public square, where stands the picturesque old court-house, the modern municipal building, the city hall, in the basement of which is the public market; the Legal Building, the newspaper offices, many stores and other business edifices. Here the electric street cars on all the lines converge. On Saturday afternoon crowds of country people congregate in the square, and the mountain wagons, cloth-covered and drawn by mules or steers, lend interest to the scene. Radiating from the square, all the streets are solidly built up with brick business blocks. On all sides of these lies the residence part of the town, built on the undulating land, not too closely, the average residence lot having a 75 foot frontage.

There are few cities in the South which have a larger number of beautiful residences. Many people who have been attracted to Asheville because of its delightful and healthful climate are making it their permanent home, and have built modern, and, in a

number of instances. luxurious homes, one of them, that of Mr. George Vanderbilt, being the most costly private residence in America. The city is amply supplied with excellent modern hotels, and there are scores of boarding houses wh[ere] comfortable accommodations may be had. The two leading hotels, the Battery Park and Kenilworth Inn,

COUNTY C[illegible] [illegible] A[illegible] [illegible]

rank high among the best resort hostelries of the country, and each has accommodations for from four to five hundred guests.

It is the peculiar climatic features of the Asheville plateau, added to its charming natural scenery, which have given this country its great reputation. These have been admirably summed up by S. Westray Battle, M. D., in an article recently published in the *Medical Record* of New York. He says:

"Nestled in the heart of the Alleghanies, cradled by the Blue Ridge and Great Smokies, stretches the Asheville plateau, a most desirable and beautiful section of country, in close touch with the East and North, and most accessible from all points South and West. It has become the great sanatorium of the eastern United States. It enjoys a climate *sui generis*, representing the golden mean of altitude and latitude and the several meteorological conditions which go to make up a wholesome and fascinating resort. Nowhere east of the Rocky Mountains is there anything approaching it to be found for fall and winter, spring and summer—an all-the-year-round retreat. It is cool in summer, yet the winters, shorn of their harshness by reason of its southern latitude, induce almost daily out-of-door exercise in the way of shooting, riding, driving or short mountain excursions on foot. For lovers of golf it is ideal; and at

[illegible] ASHEVILLE, N. C.

Asheville, the center of the plateau, are united the comforts of a city with the delights of the country.

"The plateau is an elevated tableland, somewhat triangular in shape, embracing some six thousand square miles of western North Carolina, with a general elevation of two thousand feet above the sea level, though altitudes up to six thousand feet may be had for the climbing any day in the year. Hills, valleys, rivers and forests so diversify this intramontane expanse as to make it lovely and restful to the eye beyond the power of my pen to portray.

"The mean temperature of spring is 53.49° F.; that of summer, 70.72° F.; autumn, 55.1[illegible]° F., and winter, 3[illegible].[illegible]7° F.; while for the year it is 54.1[illegible]° F.; with a mean relative humidity of but 65 per cent.

[illegible] ASHEVILLE, N. C.

"There can hardly be room for controversy that upon this plateau may be enjoyed the golden mean of [illegible] climate. [illegible] dry, [illegible] atmosphere, [illegible] clarity [illegible] in [illegible] majority of [illegible] that it [illegible] appeal to the [illegible] the [illegible] and to him of modest [illegible] to enjoy his [illegible] away [illegible] whirl of the city and [illegible] merits of the climate, or the [illegible] let me briefly [illegible] without bestowing [illegible] It is pre-eminently a suitable [illegible] stages of pulmonary phthisis [illegible] such cases as can and will get out [illegible] take the benefit of [illegible] healing qualities [illegible] Conditions which seem [illegible] propagation and prolong the species [illegible] do not exist here. Wounds [illegible] operative procedures of the gravest character are very rarely followed by septic infection.

The mortality from pulmonary phthisis is not large in any part of North Carolina, being, according to the mortality tables of the tenth census (1880), 15.4 for every [illegible] of population throughout the State. But it is interesting to note that the mountain counties show a mortality of only 10.6 in every [illegible] of population, as against [illegible] for every [illegible] of population of all the other counties of the State in the aggregate; or in other words, in a State in which pulmonary phthisis does not figure prominently in the

of the blood improved, by a sojourn at moderate elevation; above six thousand feet the appetite [illegible] diminished and the digestive organs are frequently disordered, whereas a medium altitude usually increases the desire for food and quickens digestion. By reason of its medium altitude, contra-indications to a residence upon the plateau are few, though organic disease of the heart where the circulation [illegible] much disturbed must not be lost sight of. Of cour[se] those who are in advance[d] phthisi[s] and are too feeble to breathe the out-of-door air and take some sort of out-of-door exercise are better off at home with their friends, surrounded by comforts that cannot be supplied elsewhere."

The drives round about Asheville are unex-

celled anywhere for the lovely views they offer. Here[illegible] riding is in great favor [illegible] because of the hills, bicycling is but little in[illegible] [illegible] out-of-door life, especially [illegible] is [illegible], and there are [illegible] [illegible] [illegible] [illegible] [illegible] [illegible] [illegible] will tax the endurance [illegible] of all [illegible] ve the

p[rofe]ssional Alps climber. During the spring and early summer these mountain sides are radiant in the blossoms of the laurel, the rhododendron and the azalea, and for miles along the edges of the purling Swannanoa its banks are one solid mass of these exquisite flowers. With every turn of the road a new and exquisite panorama is spread before the enraptured gaze. Peak after peak comes into view rising to majestic height, and clothed to the very summit with deep green forest. It is a matchless region, to which all others except those of the far West are incomparable.

About two miles from the heart of Asheville, and upon one of the steps of the mountain range, is Mr. George Vanderbilt's magnificent chateau, the *chef d'œuvre* of the late Richard M. Hunt's architectural creations. It was begun in [illegible] and was completed in [illegible]. The building [illegible]

[illegible] THE SKY

said to have cost upwards of $[illegible], and as much more has been expended upon the surroundings and the vast estate of a hundred thousand acres. All of the landscape gardening and the development of the park shows the master hand of Mr. Frederick Law Olmstead, under whose direction the improvements have been made. [illegible] model roads have been constructed, [illegible] hundreds of thousands of [illegible] plants and shrubs have been [illegible]. In every line of [illegible] and floriculture there [illegible] development [illegible] in order not only to [illegible], but to furnish a [illegible] which would be an influence [illegible] raising the standard of the entire [illegible] State.

The mansion [illegible] rated version [illegible] of Francis I and of the [illegible] the [illegible]. It is exceedingly [illegible] and the general effect is achieved [illegible] the employment of decorative [illegible] stood spell-bound upon the esplanade of [illegible] magnificent chateau, and looked out upon the wild [illegible] of mountains which stretch away in every direction until lost behind the curtain of the horizon [illegible] understand why Mr. Vanderbilt selected this particular spot of all others in America for [illegible] a home, which is a supreme among the [illegible] spot [illegible] the creations of Nature.

From Asheville [illegible] may turn in any [illegible] locality in the [illegible] is sublime and [illegible] lent accommodation [illegible] hotels. [illegible]

invalids [illegible] This section has [illegible] level, [illegible]

Thr[illegible] about the entire distance between Waynesv[illegible] and Murphy the country is sparsely settled, and so little affected by the inroads of modern civilization that over much of the area one may see the clear streams, the dense forests and the rugged mountains in their native wildness and beauty. Cherokee Indians have, until recent years, traversed the forests and wandered along the streams which their fathers named: the Tuckaseegee, Savannah, Tennessee, Ellijay, Cartoogachay, Taskeegee, O[illegible], Stek[illegible], Tusquittah, Nantahala and others. It is a wilderness as sublimely beautiful as it is solitary and grand, an elysium for the health-[illegible], a paradise for the sportsman.

During long ages of the past these streams have been carving deeper and deeper their channels between the mountains. So slowly has the work progressed, and so [illegible] has the vegetation grown, that everywhere from the mountain tops to the banks of the streams the [illegible] trees, shrubs and flowers.

The gorge of the Nantahala River, through which the [illegible] more than a dozen miles, is by [illegible] most picturesque and beautiful [illegible] North [illegible] of the title-page illus[illegible] the upper end of this gorge, looking

northeast down the river. On the right Cliff Ridge rises almost perpendicular to nearly 2,000 feet above the river; while on the left the spurs of the Great Smoky Mountains rise nearly as high and are nearly as steep. Between their bases the gorge is so narrow that in many places there is hardly space enough for both the railroad and the river. Talc and marble abound in these rocky, forest-covered slopes, and the lumbering and timber interests at Dillsboro, Bryson City and other points are extensive.

On the line of the Southern Railway between Asheville and Spartanburg, S. C., there is a beautiful and picturesque region which has long been extremely popular with tourists, and in which there are numerous resorts well patronized both in summer and winter. The nearest of these resorts to Asheville is Skyland, an attractive little place nestling down close under the protection of the nearby

[illegible]

mountains. Beyond is Henderson, twenty miles from Asheville, and located in full view of the mountain peaks of Tryon, Little Hog Back, Glassy, Pinnacle, Cæsar's Head, Hebron, Hog Back, Pisgah, Bushy, Craggy, Black, Hooper's, Bear Wallow, Sugar Loaf, Chimney Rock, the Shaking Bald and Point Lookout, which, rising above the plateau, form a complete panorama and amphitheatre, making the view from the town grand and majestic beyond comparison.

Three miles beyond Henderson is Flat Rock, one of the most charming little resorts in western North Carolina, and [illegible] spot where many prominent people from Southern cities spend the summer months. It is in appearance a little corner of old England tucked down in this North Carolina paradise.

W[illegible] only a quiet little hamlet of [illegible], Flat Rock was discovered [illegible] years ago and taken possession of by a company of French and [illegible] gentlemen who owned lands in

WAYNESVILLE, N. C., AND THE BALSAM MOUNTAINS

South Carolina and Georgia. Among the very first were the Count de Choiseul, the Barings, the British consul Mollyneaux, and a half dozen or so of planters and their families from the coast, who, finding this climate so entirely different from their own, the place so unique of all others in the mountains, set up their summer lodges here. At the present time these places, about fifty in number, cover an area of five or six miles and are picturesquely wooded with the fragrant pine, the oak, the hickory, dogwood, sassafras, the crimson maple, the hemlock and the holly, thickly interspersed with the beautiful mountain laurel and azaleas of colors the most gorgeous and the most delicate, while flowers and ferns fringe with beauty "the banks and braes and streams around." Streams flow into artificial lakes shut in by rugged hills, and beautiful with blue inverted skies.

... WHITE SULPHUR SPRINGS, WAYNESVILLE, N. C.

Paths winding through the sweet shades lead out upon different points of interest, among them the quaint and picturesque church built eighty years ago by the Barings, of London banking fame, and called "St. John in the Wilderness." Its nearby vine-covered rectory is so classical that one involuntarily expects to see the Vicar of Wakefield step out from its portals. Flat Rock is provided with ample accommodations for the entertainment of guests, and no more restful or healthy spot exists on the American continent. All of the neighboring old estates are thrown together in such a way that visitors, who are always welcome to do so, may enjoy many miles of beautiful drives from which the loveliest of mountain and nearby views may be enjoyed.

Saluda, nine miles beyond Flat Rock, is 2,250 feet above sea level. In approaching it from the south there is for three miles an ascent of 237 feet to the mile, two locomotives being necessary on each train. The little town, handsomely situated on this elevated plateau, is nestled amid forest-covered hills ranging from [illegible] to [illegible] feet above the depot. On these hills, families, principally from Columbia, Charleston and the coast, have built their airy, shady homes, and spend their summers enjoying, in the cool breezes of the mountains, the repose of country life without its loneliness.

Saluda has two good hotels and several boarding houses, so that the stranger within its gates is certain to find accommodations of a satisfactory type.

The little town of Tryon is forty-three miles from Asheville and twenty-seven from Spartanburg, and is 1,500 feet above sea level. The scenery hereabouts is beautiful beyond description. The mountains are covered to their very summits with verdure, and whether in the budding of spring, the full foliage of summer, or the gorgeous coloring of autumn, the ever-changing picture is always one of beauty, charming to the eye. The waterfalls and cascades of the Pacolet River and its tributaries are far famed. The Horseshoe Falls, on Spring Mountain, tumble down the mountain side a distance of 350 feet. A good road leads to the top of Rocky Spur, a peak 4,000 feet high, a trip that can be made between breakfast and dinner; and the sightseer will find a comfortable hotel, the Skyuka, near the top of Tryon Mountain, passing

[illegible] is a very attractive mountain town. It enjoys [illegible] from the rich agri- [illegible] It also has ex- [illegible] the largest [illegible] The [illegible] between [illegible] by the graceful [illegible]

[illegible] over near the [illegible] one of the most [illegible] on the [illegible] can be compared [illegible] Furca and other famous passes [illegible] miles between [illegible] only half of this [illegible] gorges. [illegible] mountain [illegible] every turn there [illegible] entrancing [illegible] the [illegible] lower [illegible] where you [illegible] crystal streams [illegible] down the mountain [illegible] towering summit [illegible] Mountain, [illegible] the profile of an old man's face. The road winds among the hills for six or seven miles and then strikes the Yadkin River, and [illegible] spring which is [illegible] The [illegible] [illegible] we climb [illegible] growth which [illegible] [illegible] the [illegible] a tangle. The [illegible] ascent, and turning [illegible] in the road, the whole [illegible] were, lies below us. [illegible] down into the [illegible] John's River, [illegible] the sweeping [illegible] as they rise [illegible] seem to melt into the blue of the [illegible] has been made [illegible] [illegible] made among the [illegible] distant peak, [illegible] with [illegible] [illegible] [illegible] [illegible] the mountains and [illegible] with silver. You [illegible] [illegible] There [illegible] [illegible]

Here summer reigns with no fierce sway, during [illegible] the highest temperature [illegible] [illegible] maximum range [illegible] from [illegible]

[illegible]

strikes against the face of the cliff. The air [illegible] over the top, and the force with [illegible] gives the name of Blowing Rock to the [illegible]. When the winds are right any light [illegible] hat or fresh thrown from the [illegible] bottom thousands of feet [illegible] upward and back again to the spot [illegible]. The name of the cliff has [illegible] nearly where the road to Boone [illegible]

R[illegible] cosmopolitan. There [illegible] and dance and ride to the heart's desire [illegible] may [illegible] [illegible] trunk and go [illegible] from sun[illegible] [illegible] invalids [illegible] [illegible] sensation [illegible]

where that good old mother takes her children into her lap and soothes their jangled nerves; where the doctors are never in evidence and the medicines always delightful."

The twenty-mile drive from Blowing Rock to Linnville is over a road which for surveyors' skill has no superior in America. For miles it traverses the forest primeval, and from one point furnishes a view of matchless grandeur, and from another a glimpse of some sweet, quiet valley with perchance the modest home of some mountaineer and its little clearing far away below. Everywhere the wild flowers grow in profusion, and countless mountain streams murmur greetings as you pass. Nine and a half miles beyond Blowing Rock the traveler comes to the eastern boundary of the great park of 16,000 acres owned by the Linnville Improvement Company, and the first view, from a point 1,500 feet above it, is had of the beautiful valley of the Linnville River. Far away to the west and nestling in the heart of the valley is the charming Esceola Inn, with its surrounding picturesque cottages. You may imagine it a little bit of Switzerland dropped down in our own "Land of the Sky." The inn is of pleasing architecture, and has all the conveniences found in the best resort hotels, including an excellent orchestra. Nearby is a large artificial lake, and in every direction the beautiful roads tempt the visitor to ride and drive, while those who enjoy

hunting [illegible] and in the neighboring streams opportunities which will afford a fruitful basis for [illegible] in the days to come. From the Esceola Inn [illegible] road continues to Cranberry, and from [illegible] road may be taken to Roan Mountain, [illegible] to Knoxville.

[illegible] section of the "Land of the Sky" which [illegible] of its solitary and grand [illegible] Hickory Nut Gap and Chimney Rock [illegible] from Asheville, and may be [illegible] by mountain conveyance. The [illegible] pay the tourist, for the rugged [illegible] and the deep [illegible] gorges are [illegible] [illegible] hotel at Chimney R[illegible] comfortable accommodations.

H[illegible] French Broad River, is the best-[illegible] town [illegible] North Carolina after Asheville. It is [illegible] of the latter place and but a short [illegible] state line. It [illegible]

Asheville by many [illegible] was long [illegible] days [illegible]

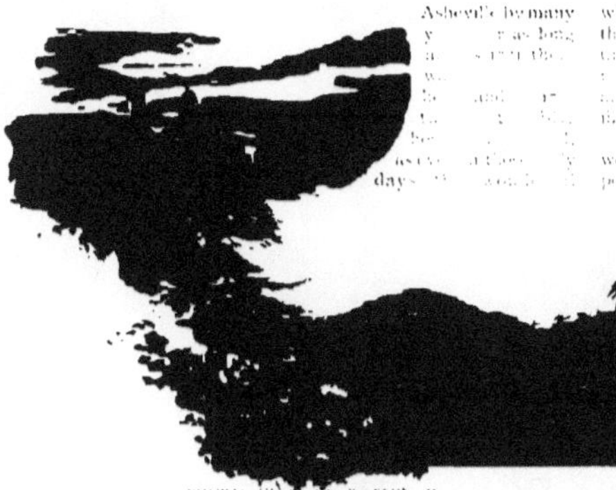

curative properties of its waters had become known, and the settlers for hundreds of miles around were wont to bring their sick here for the benefits to be derived from the baths. To-day, the Hot Springs of North Carolina is one of the best-known health resorts in America, and its handsome modern hotel, the Mountain Park, is frequently taxed to its utmost capacity to accommodate the great number of representative people who gather here from North, East, South and West.

The Southern Railway, from Asheville to the Hot Springs, follows closely for the entire distance the tortuous windings of the historic and beautiful French Broad River.

Of such a stream the poets might sing, for it is matchless in its setting of mountains and in the beauty of its graceful curves. Nearly the springs the mountains, as if determined to hold it [illegible] to the lowlands, crowd in the closer and [illegible] high and rugged barriers on either [illegible] between which the river, lashed into a fury of foam, tosses and [illegible] as if impetuous at the encroachment [illegible] mighty [illegible] victory [illegible] [illegible] meanders [illegible] toral valley. Hot Springs [illegible]

which are located the modern bathing houses where the hot baths may be taken in handsome marble [illegible] The hotel has [illegible] [illegible] have wrought many [illegible] wonder [illegible] The surrounding country has many points of interest, notably the famous and historic Paint Rock, which marks the dividing line between North Carolina and Tennessee, and [illegible] from the hotel.

In the foregoing pages there has been given a brief outline of what the noble State of North Carolina has been accomplishing in the line of material development and a suggestion of its innumerable and charming natural features.

That it is destined to become a great commercial and industrial empire there can be no question, and that it will eventually be to the United States, in a tourist sense, what Switzerland is to Europe, there is but little doubt in the minds of those who are familiar with the great advantages it possesses for both scenery and health. The fast and perfect train service of the Southern Railway has brought it to a neighborly distance of the chief centers of population in the Eastern, Southern and Central States. Its citizens are alert, intelligent and enterprising, and its undeveloped opportunities invite the farmer and mechanic, the banker and the merchant, the manufacturer and the artist to a share in the distribution of its wealth.

The growth of the South in all lines of human activity is [illegible] In the [illegible] State of North Carolina [illegible]

South Carolina

TO South Carolina belongs the high honor of being one of the earliest, if not the very first colony to offer a premium for immigration. This stroke of enterprise was made over two centuries ago, in 1670, when the low-rent inducement was held out by the Ashley River settlement, under Sayle, of land at halfpenny per acre for five years. This invitation wafted across the sea brought many settlers to the palmetto-fringed State, and marked the beginning of a progressive policy that is being followed, in this later time, throughout all the South with the most prosperous and beneficent results.

To-day South Carolina remains true to her past. She offers a comfortable home with all the conveniences of modern civilization, fine farming land at a nominal price, good titles to every foot of it, and a cordial welcome to the home-seeker. To capital she offers fair and just laws, ample protection to property, an honest and honorable class of working people, good markets at home and the best facilities for reaching those abroad, abundant and safe banking facilities, in many instances exemption from local taxation and a helping hand—a hand with dollars in it.

The inheritance of enterprise in fostering immigration is not the only bequest from the past of which South Carolina is proud. Her patriotic record during the War for Independence is a splendid legacy of deeds of high emprise, all of which made for liberty. Her Laurenses, her Rutledges, her Pinckneys were noble contributors to the cause of the country's freedom. They stood with the Washingtons, the Jeffersons, the Henrys and the other immortals of the "Old Dominion," and won for South Carolina the high place of being second only to Virginia, among the Southern colonies, in the heroic struggle to break the bonds of England.

It was from the friendly cover of her forests that Marion and his men darted and struck telling blows for freedom. It was at Cowpens, within her borders, that Colonel Washington defeated the brilliant English cavalry leader Tarleton, and made the occasion for one of the choicest *bon mots* of patriotism. In a London drawing-room, years after the Revolution, it is related that Colonel Tarleton was recounting his exploits in the Lower Carolina. On referring to the battle fought at Cowpens, a noble lady inquired if it was not there that he had met Colonel Washington. Tarleton replied that it was, and added, in a contemptuous way, that the American was an illiterate rowdy of a soldier. "Ah, my dear Colonel," the lady is said to have responded, as she looked at Tarleton's fingerless hand where Washington's sword had struck, "though he may not have been able to write, he certainly could make his mark."

In the first foreign difficulty to confront the republic, the controversy with France in Washington's administration, it was a son of traditions of patriotism and of enterprise which run through the whole of the State's history. They are an inspiration to-day in the work to achieve a high place for South Carolina in America's industrial progress. That they form an effective inspiration, these pages, giving a brief chronicle of [illegible] accomplishment, will serve to show.

As a preface to the State's resources and their development, it will be interesting to look for a moment at South Carolina's topography. The State is naturally divided into three parts, the hill or up-country, the middle country, and the coast or low country. The latter is rich in vast timber tracts and in boundless swamps full of the glory of cypress and pine. Here grow the palmetto, which gives the State its name, and the long staple cotton, fine as [illegible] silks of the Orient. This is a land of sunshine, whose flora is radiantly beautiful the whole year through. It is a land where the mellow tints of the long ago lend a dreamy charm to life, and make the past almost as attractive as the future.

The middle country is undulating, broken here and there by sand hills. Dotted over it are beautiful farms and busy towns. The rivers, on their way to the sea, widen out and become the water highways for a considerable commerce. It is a region of prosperous agriculture, and the home of a progressive people.

[illegible]

But it is the up-country which seems most favored, as was the hill country of Judea, where "shepherds tended their flocks by night." It is not only a land of promise, but, in the happy phrase of the region, "a land of fulfillment" as well. A fertile soil yields a bounteous variety of crops, and a busy industry, to the hum of a million spindles, changes raw material into finished product. In this section are some of the most progressive cities in the South. It is the rich Piedmont, whose opulent stretches have been followed thither from their beginning in Virginia through North Carolina.

[illegible] HOUSE, COLUMBIA, S. C.

The scenery of this region is picturesquely beautiful and presents many attractions to both health-seekers and tourists, as well as to settlers. The Saluda Mountains, which constitute a portion of the northwest boundary, and which [illegible] spurs of the Blue Ridge, have several peaks which rise like turrets on a battlement, reaching a height in King's Mountain of 1,692 feet, Paris Mountain, 2,054; Table Rock, 3,000; Cæsar's Head, 3,118, and Pinnacle Mountain, [illegible] feet, and form the background to a most delightful landscape. In the Piedmont are many rapids and falls affording excellent water power, and there are numerous points admirably located for mill sites and a variety of manufacturing plants.

South Carolina is one of the leading States of the South, both in the production and manufacture of cotton. She raises annually nearly one-tenth of the American crop. This great staple, always as good as gold, has the first place in the State's agricultural products. In [illegible] the crop was [illegible] bales, and it had a

total value of $[illegible]. In 1896 the crop brought $27,[illegible],700, and in 1897 $[illegible].

For this great crop there is a home manufacturing market that is every year consuming a larger proportion of the total product. The South Carolina cotton mills in 1896 consumed 254,[illegible] bales, and in 1897 327,643 bales, nearly one-half of the State's 1897 crop. As home consumption saves the cost of transportation, it means, of course, an increased profit to the producer. There are at present 98 mills in the State, with 1,250,324 spindles. Besides all the general causes for drawing the cotton mills from North to South, South Carolina has the special cause of abundant and cheap water power. As showing what a great advantage this is, the average cost for the whole State of one water horse-power is $1.7[illegible], while the cost for the same power in Lawrence, Mass., is $14.12; in Lowell, $2[illegible]; Paterson, N. J., $37.50; Cohoes, N. Y., $20, and in Dayton, Ohio, $37.

This conjunction in South Carolina of raw material and natural power has led to a development of cotton manufacturing that is unprecedented in American industrial history. In 18[illegible] the United States Census showed that South Carolina had 14 cotton mills of 82,334 spindles, employing 2,0[illegible]3 hands, paying $[illegible] in wages, having $2,77[illegible] capital, consuming 15,[illegible] pounds of cotton, costing $1,[illegible], and yielding a product valued at $2,[illegible].

The statistics of cotton mills for the year ending August 31, 1897, made by Henry G. Hester, secretary of the New Orleans Cotton Exchange, show that South Carolina had a total of 1,250,324 cotton spindles, of which 1,05[illegible],[illegible]24 were in operation, 17,[illegible] were new and just starting up, 1[illegible]2,000 were new, not completed, and only 15,[illegible] were idle. Such a striking exhibit tells a most eloquent story of progress.

A growing home market offered by these mills is but one of the hopeful factors in the future of cotton raising. The other is a more intensive method of farming. The latter has been attended by marked increase in yield, and has made it clear that cotton can be raised at a profit when its price is low.

But S[illegible] Carolina is a land of [illegible] as well as of [illegible] In [illegible] a farmer [illegible] the State [illegible]

was revived. The crop has been found so profitable that it will, without doubt, be produced in larger quantities each year. The mahogany and gold leaf grow equally well, and the quality is so high that South Carolina tobacco commands top prices. The handling of the

[illegible]

[illegible] grown to a greater or less degree [illegible] one to two million ba[illegible] they grow well, but the yi[illegible] a very [illegible] crop. They are [illegible]

crop has made profitable the opening of warehouses for the sale of [illegible] tobacco in many of the towns, thus contributing [illegible] to both business and agricultural progress. [illegible] a product so small as to be scarcely reckoned [illegible] among the State's resources, tobacco has advanced [illegible] an annual crop of several million pounds, and to a place of great importance on the right side of South Carolina's annual balance sheet. It grows in all [illegible] of the State, and thrives especially in the fertile Piedmont region.

[illegible] has always been raised for home consumption, but the possibilities of its cultivation for profit have only recently received the attention they merited. The energetic farmers and fruit growers of this section have [illegible] the peculiar adaptability of the soil of this region, the even temperate climate, and the rapidly [illegible] demand of the Southern, Northern and [illegible] added to the now excellent facilities in [illegible] have opened a broad field in [illegible] growing, and one in which there are [illegible] profits. Now, on the beautiful hill[illegible] the Piedmont clustering vineyards are offering [illegible] fruit to the wine press, and bringing to their [illegible] a handsome return. In Oconee, Greenville, [illegible] Richland counties in particular, wine[illegible] found profitable. Both the grape and [illegible] to South Carolina [illegible] having been

found growing wild on St. Helena Island off the coast, by the first settlers. The peach, [illegible] apple [illegible] as well [illegible] the [illegible] famous for its fruits and its crops through-out its entire [illegible] The [illegible] without [illegible] losing any [illegible] their delicate flavor. As

[illegible]

many things to be superior to jute. The consumptive [illegible] peace and ease [illegible] the air [illegible] with the exhalations of the royal pine. The worker in wood finds in it as beautiful paneling as the artistic heart can desire."

Lumber is manufactured into doors, sash and blinds, wagons and other articles of commerce at Abbeville, Aiken, Anderson, Columbia, Greenville, Greenwood, Newberry and other towns. As showing how much this industry can be developed, of the State's total area of [illegible] acres, 12,[illegible] are covered with timber. Besides pine, there are the magnolia, the sweet and black gum, black walnut, cypress, elm, hickory, maple, sycamore, ash, chestnut, beech, locust, persimmon, dogwood and poplar.

This inexhaustible supply offers many opportunities for profitable investment, and is an illustration, on a far-reaching scale, of the prodigality with which Nature has dowered South Carolina with resources. Nothing is lacking to complete a round of native opulence—there are the cereals and fruits in abundance for bodily sustenance, the silky cotton for raiment, rivers swiftly coursing with power for factories, forests lifting up in their tall trunks many a length of stout timber for house and church and school. To man's hand Nature [illegible] being everything he needs, and [illegible] climate.

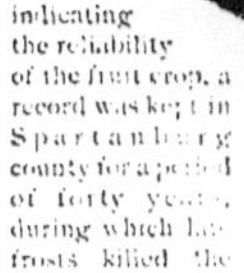

indicating the reliability of the fruit crop, a record was kept in Spartanburg county for a period of forty years, during which late frosts killed the fruit but once. This reliability is a characteristic of the State. South Carolina is in truth "the land of fulfillment."

As might be expected from a State having the palmetto as a sign manual, South Carolina is rich in timber. There is invested in the lumber industry over $4,[illegible], and the annual [illegible] value of $10,[illegible]. There are nearly [illegible] lumber mills in the State, scattered in every section, and all operated at a profit. There are not less than [illegible] acres of pine furnishing a [illegible] [illegible] A [illegible] recently said: "[illegible] in the South Carolina pine [illegible] is [illegible] [illegible] an inflammatory [illegible] [illegible] paints, in [illegible] In [illegible] [illegible] [illegible] [illegible] [illegible] [illegible] and proved in

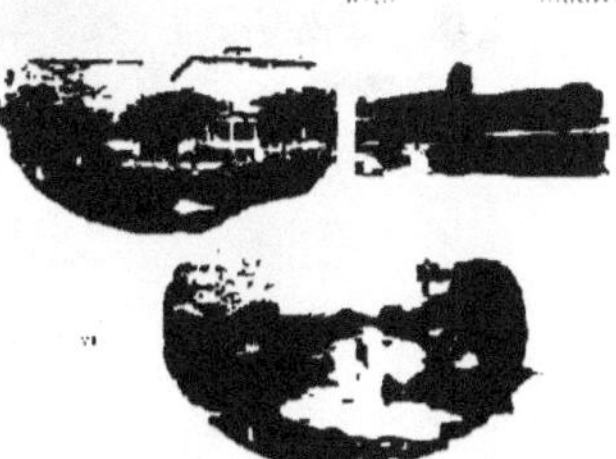

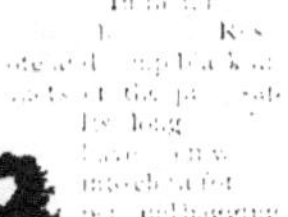

As a building material, in addition to lumber, th[illegible] stone in abundance [illegible] South Carolina gran[illegible] It [illegible] Railway, [illegible]

[illegible] is estimated, [illegible] the stupendous [illegible] While there is still con[illegible] North [illegible]

section of South Carolina. On this main [illegible] are Blacksburg, Gaffney, Cowpens, Clifton, Spartanburg, Wellford, Greers, Greenville, Easley, [illegible] Central, Calhoun, Seneca and Westmin[illegible] At Spartanburg, [illegible] with that [illegible] west through [illegible] Columbia, where it [illegible] which leaves [illegible] at Charlotte, N. C., and runs [illegible] Fort Mill, [illegible] Winnsboro, [illegible] and Ridgeway. Between [illegible] Columbia are the towns of [illegible] Rich Hill, [illegible] Union, Strothers, [illegible] noted for their [illegible]

From Greenville, [illegible] main stem south [illegible] Columbia [illegible] Piedmont, [illegible] (from [illegible] branch [illegible] to [illegible]

Columbia con[illegible] made with the Florida Central & Peninsular upon which the through trains of the Southern Railway continue on to Savannah, Everett, Jacksonville and Florida points.

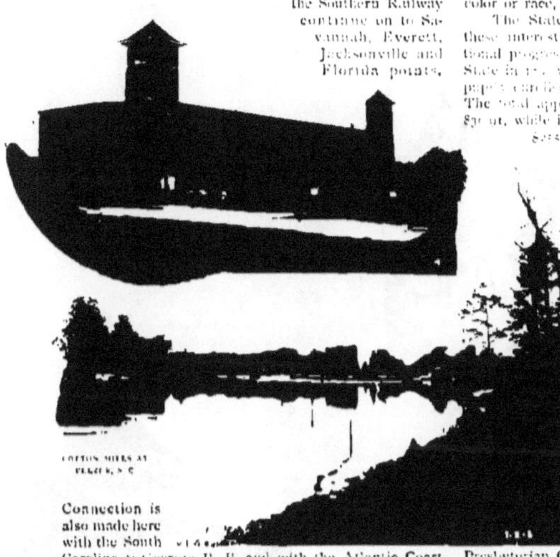

Cotton Mills at [illegible], S. C.

Connection is also made here with the South Carolina & Georgia R. R. and with the Atlantic Coast Line. The through trains of the Southern between Charleston and Asheville, N. C., are operated over the former road between Charleston and Columbia. From Columbia the Southern Railway runs southwesterly to Aiken, S. C., and Augusta, Ga., passing through the towns of Lexington, Leesville, Batesburg, Ridge Springs, Johnstons, Trenton, Vaucluse, Graniteville, King and Bath. This is known as the region of cotton mills, and all along this portion of the line there is great development in the cotton milling industry.

But the State, while enlarging her commerce, is not forgetful of her schools. Her constitution provides for a two-mill tax on all property, and a one-dollar poll tax on all men between twenty-one and sixty years of age, for the support of her public school system. The zeal for education is not of recent growth. In 171[illegible] South Carolina [illegible] her first free schools, and a [illegible] system covering the State was [illegible] in [illegible] and reorganized in [illegible]. This system provides for the free instruction of all children between the ages of six and sixteen irrespective of color or race, in the primary and intermediate grades.

The State Superintendent of Education supplies these interesting statistics of South Carolina's educational progress. The number of public schools in the State in [illegible] was [illegible]; in [illegible], [illegible]. The number of pupils enrolled in [illegible] was [illegible]; in [illegible], [illegible]. The total appropriation for schools in [illegible] was [illegible], while in [illegible] it was [illegible], an increase of [illegible].

The State University has one branch, the South Carolina College, at Columbia; another branch, the State Military School at Charleston; and still another, Clemson College for the Instruction of White Males in Agriculture and the Mechanical Arts, at Calhoun. Winthrop College, for young women, at Rockhill, and the Institute for the Education of the Deaf and Dumb, are also under State patronage. In addition to these State institutions there are many private and denominational colleges and schools, such as Furman University at Greenville, Wofford College and Converse College at Spartanburg, Claflin University for Colored Youth at Orangeburg, Medical College of South Carolina at Charleston, College of Charleston at Charleston, Greenville College for women at Greenville, Leesville College at Leesville, Female College and Erskine College for young men at Due West, Presbyterian College of South Carolina and Thornwell Orphanage at Clinton, Newberry College at Newberry, Presbyterian College for women, Allen University and Benedict College at Columbia, Cooper Limestone Institute at Gaffney, Sumter Institute at Sumter, Clifford

[illegible]

[illegible] college for young women at [illegible]

[illegible] South Carolina it would be

[illegible] indeed [illegible]

in prosperity. When a city has a dozen strong points, it offers convincing assurance of a splendid destiny. Such a city in truth is Columbia. Being the state capital, it is the center of political influence. The legislature meets annually in January, the session lasting from [illegible] forty days. It is the railroad center [illegible] State. Eight lines radiate from its hub [illegible] great Southern Railway in four directions [illegible] to Charlotte and the North, to Augusta and [illegible] South, to Spartanburg, Asheville and the West, [illegible] Greenville, Anderson and all points in the Piedmont. By reason of these unsurpassed railroad facilities and its mid-state location, Columbia is a great distributing center. It is [illegible] connected with and almost equidistant from the ports of Savannah, Port Royal, Charleston and Georgetown. Its wholesale and jobbing business, already considerable, is increasing steadily because of advantageous rates and the facilities for quick distribution.

Columbia is one of the educational centers of the State. It is the seat of the South Carolina College, the State institution of highest learning; the Columbia Female College, Methodist; the Presbyterian College for women; the Presbyterian Theological Seminary; [illegible] South- [illegible] Allen University [illegible] Benedict Institute [illegible] Columbia Business College [illegible] the Chicora College [illegible] Institute, three [illegible] school of stenography and typewriting [illegible] public school system, of ten grades, for [illegible]. Two great State institutions, the hospital for the insane and the penitentiary, are located here.

The city has extensive manufacturing [illegible]. The Columbia canal, [illegible] and one-half miles long, yields [illegible] horse-power. Ten thousand horse-power of [illegible] is now electrically developed by [illegible] owners of the property, the Columbia Water Power Company, a corporation of New England capitalists. The [illegible] electric powerhouse, with its [illegible] horse-power [illegible] is one of the sights of the [illegible] progressive [illegible] came from [illegible] the South. This electric power [illegible] near the city, and is the

The three [illegible] cotton mills in C[illegible] operated by [illegible]

[illegible]

little bad weather from September to May. The air is [illegible]

[illegible] libraries, two club houses [illegible] public two daily newspapers, an opera house, hospitals for white and colored patients, a fire department, over eleven miles of electric railway, and all the [illegible] of modern city life.

[illegible] exceedingly good. During the [illegible] year the death rate was only [illegible] per thousand [illegible] The water [illegible] by [illegible] Jewell [illegible] Johns Hopkins University [illegible]

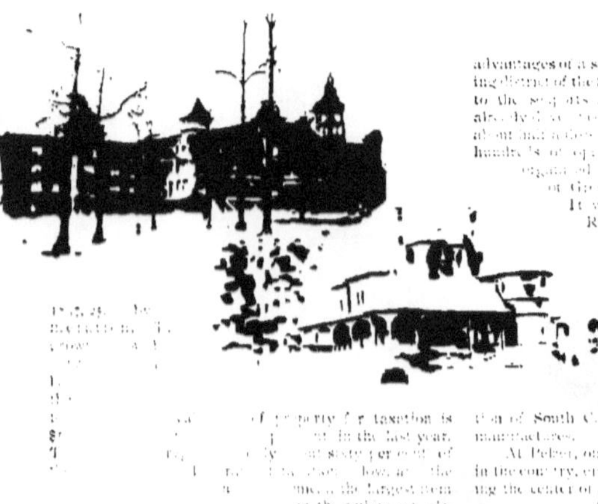

[illegible] of property for taxation is [illegible] in the last year. [illegible] about sixty per cent of [illegible] the largest item [illegible] of the public schools.

advantages of a situation in the heart of the manufacturing district of the South, excellent transportation facilities to the seaports and large markets, and water power already developed in the middle of the city. There are about [illegible] cotton mills in Greenville, employing hundreds of operatives. These companies have been organized and are largely capitalized by citizens of Greenville.

It was at Piedmont, on the Saluda River, directly south of Greenville, that the first cotton mill in the upper part of the State was erected. This was in 187[illegible], and the mill is still in successful operation, having been often enlarged to meet the demands of a growing trade. The example of its success was contagious, and there have clustered about it, in the quarter of a century since its erection, mills by the score, till this section of South Carolina leads all the South in cotton manufactures.

At Pelzer, on the Saluda, is one of the largest mills in the country, employing [illegible] operatives, and constituting the center of a prosperous industrial community.

The city of Spartanburg, located in the largest cotton manufacturing county of the South, has a population of over [illegible]. Its altitude is [illegible] feet above sea level, affording cool breezes even in midsummer. There are twenty-two separate mills in Spartanburg County, operating [illegible] spindles, and employing [illegible] operatives. For cotton and wages nearly $[illegible] is paid out annually, and [illegible] bales of cotton are consumed.

Besides eight graded schools, Spartanburg has two institutions of higher learning. One of them, Converse College for young women, has a high standard and good equipment. With a faculty of thirty, commodious buildings, and a campus of [illegible] acres, it has in [illegible] an enrollment of [illegible] over [illegible] students, [illegible] one hundred per cent. over the preceding year. The other, Wofford College, is one of the oldest in the State, and is under the control of the Southern Methodist Church. It has about fifteen buildings and one hundred and fifty students. Agriculturally, as well as industrially, Spartanburg County is rich, having a fertile soil, adapted to grains and fruit.

Abbeville, a flourishing town in the "up-county," has a population of about 3,500. Its location on high and undulating ground gives it excellent drainage, and its climate is temperate winter and summer. The town has adequate public schools, two colleges for colored students and some nine or ten churches. A large cotton mill has lately been erected, operating 10,000 spindles and manufacturing brown homespuns. There is also a cotton-seed oil mill, and the usual industries of a thriving town find representation. A productive soil and many streams capable of supplying abundant water power make this section of the northwestern part of the State an attractive one for investors and others seeking business openings.

The city of Anderson, north of Abbeville, in the county of the same name, one of the richest and most progressive in the State, has an energetic population of more than 6,000. It is in the heart of a most excellent agricultural country not far from the Blue Ridge Mountains. Anderson has developed by natural, healthy growth. Nearly everyone owns his house, be it large or small. The various religious denominations are well represented and the schools are excellent. The Patrick Military Institute is doing good work in the education of young men. Citizens of Anderson in 1890 demonstrated their confidence in their town, as well as their business sagacity, by the erection of a large cotton mill, whose capital stock and capacity have been increased from time to time until there are now 36,000 spindles and a capital of $[illegible]. Steady employment is given 700 operatives. Two cotton-seed oil mills are kept running night and day during the season.

Greenwood, a short distance northeast of Anderson, is both a health resort and a manufacturing center. It has a salubrious climate and some notable health springs of chalybeate and lithia water. Greenwood contains two mills for the manufacture of cotton goods of fine quality, and has also an elaborately equipped oil mill. There are public and private schools, and the Brewer Normal Institute (colored). Outside the town is an extensive granite quarry.

Aiken, southwest of Columbia, near the Georgia line, is a city of 4,000 people in the sand ridge section of the State. It has become noted as a health center, and is one of the leading resorts of the South; many people whose first idea was of a merely temporary stay have become permanent residents and engaged in business enterprises. A large number of wealthy Northerners have bought property here and have built handsome residences, while all the houses that can be rented furnished are taken by this class of people for the winter. There have been at least $75,000 worth of improvements in the line of buildings and residences during the past year alone. Within five or six miles of Aiken, and within [illegible] of the county, [illegible] four cotton factories [illegible] business. The city itself [illegible] and has made [illegible] It has recently been [illegible] New York [illegible] composed of eighteen [illegible] of the [illegible] New York. The [illegible] to 15 [illegible] [illegible] The woods [illegible] are well stocked with [illegible] The [illegible] temperature of Aiken

predominated in the town's early settlers, modified by American characteristics. Of the two largest factories here, one is for cotton spinning, while the second manufactures ginghams of a fine texture for Northern and foreign markets. There are also yarn mills, employing about [illegible] hands, and a large factory in which improved machinery and farm implements are made. Much attention is given to education. Chester was the second city in the South to establish a public school system, and it is said to have the best equipped school building for white children in the State. This was completed in 1891 at a cost of [illegible]. The [illegible] Institute, comprising several large buildings, situated on a hill overlooking the country for miles around, is devoted to the instruction of colored children and youth, and to the training of teachers for the colored public schools. The city has a fine situation [illegible] feet above sea level, with good sewerage and artesian water. The country round about is fertile, cotton being the chief crop, with corn and small grains following. The county is rich in mineral deposits, and the water power available is unsurpassed.

Rock Hill, just north of Chester, is the largest town in York county, in the extreme northern portion of the State. Here is located Winthrop Female College. The present population is [illegible]; taxable value of property, $1,2[illegible]; its yearly business amounts to [illegible]; it handles annually, on an average, 18,000 bales of cotton; it has in operation three cotton factories (spinning and weaving), representing a capital of [illegible], and two more in process of erection which will employ a capital of $125,000, aggregating in cotton manufacturing a capital of [illegible]; a buggy, carriage and wagon factory, with capital of [illegible]; a tobacco factory, with capital of [illegible]; a door, sash and blind factory; a canning factory; an electric light plant, with arc and incandescent lights; a town site company; a street railway and water works company and machine shops. The weekly payroll of Rock Hill's manufactories amounts to more than [illegible].

[illegible], in the celebrated Piedmont belt, is near the Broad River, which has been called the Merrimac of the South. The looms of Lowell and of Manchester could be kept in lively motion the year round by the water of this beautiful stream. In 1875 [illegible] numbered [illegible] souls; in [illegible] population had reached [illegible]. The reasons for [illegible] found in the town's advantageous [illegible] and enterprise of its citizens. [illegible] of cotton are sold annually from wagons in the town, and the grades are superior. One of the most [illegible] mill in the South was established

[illegible] during the [illegible]

at this point in [illegible], with capital stock of $[illegible], but has since been increased to $[illegible]. The [illegible] is well adapted [illegible] spinning [illegible].

In the [illegible] buildings, [illegible] private, [illegible] make an excellent showing. It [illegible] commercial [illegible] land [illegible] a town which [illegible] about [illegible]. The Cooper Limestone Institute, named in honor of the philanthropist, Peter Cooper of New York, who bequeathed it to the Spartanburg Baptist Association, offers [illegible] young women. Its [illegible] widely known Limestone [illegible] features of the city. [illegible] is the Gaffney Manufacturing [illegible] which occupies a commanding [illegible] of the town.

Union, with a population [illegible] the main line of the Southern R[illegible] Piedmont counties. The surrounding [illegible] scenery and in natural advantages, [illegible] land and splendid water power. The [illegible] greatly interested in the [illegible] much progress during [illegible] now several cotton mills [illegible] and [illegible] looms, which [illegible] number of hands. Building [illegible] are going on, school and [illegible] the times, and altogether [illegible] lively present and an [illegible].

Newberry is in [illegible] the Piedmont and [illegible]

[illegible]

Stock-raising has proved profitable. The city has a population [illegible], and contains prosperous cotton and oil mills in addition to the ordinary industries. The city's water supply is from an artesian well and the [illegible]. An [illegible] it is claimed [illegible] manufac[illegible] which [illegible] Milwaukee [illegible]. In addition [illegible], Newberry is the seat [illegible] College.

[illegible] State [illegible] views are to be [illegible] North. Fruit [illegible] cotton and tobacco [illegible] quantities [illegible] country [illegible] town [illegible] cotton [illegible] mills.

all its own, and retains many of the characteristics of its early Huguenot days. Considered from an industrial and commercial standpoint Charleston is one of the progressive cities of the South. Its trade and commerce during the year ended August 31, [illegible], amounted to $[illegible], [illegible] over [illegible] previous year of [illegible]. There was a gain [illegible] [illegible] [illegible] [illegible] [illegible] [illegible] sale and retail trade, [illegible] [illegible] and [illegible] in phosphate [illegible]. The city's financial condition is excellent, real estate [illegible] steadily advanced, and with its new elevator and the completion of the jetties, giving [illegible] feet [illegible], and the opening [illegible] line of steamers to Europe, it is safe to reckon that the business of the coming year will reach the [illegible] mark. The phosphate industry of South Carolina, which largely centers in Charleston, is one of the State's many prosperous industries, and in its mining, shipping and handling there are millions of dollars invested.

The [illegible] sketches of the towns and cities of South Carolina serve at least to show the State's present dominating spirit, the controlling purpose, everywhere apparent, to develop manufactures. With clear appreciation of her marvelous advantages in cotton manufacturing, South Carolina is forging to the front in this important industry with a swiftness that can find no precedent. South [illegible] conquests, and the dawning century may see her sign-manual changed from the palmetto

[illegible]

GEORGIA

THE site of Chicago was bought from the Indians for less than the price of a high-grade bicycle. It does not appear from the records that the whole State of Georgia, the Empire State of the South, cost so much. "How did you get your land?" asked a newcomer of a scion of one of the old families of Kentucky. "From my father," was the reply. "How did he get it?" "From his father." "And how did he get it?" "From his father." "And how did he get it?" "Fought for it." "Pull off your coat!"

Oglethorpe was more fortunate. He did not have either to fight for it or to buy it. The shrewd commercial spirit was then lacking in the native American. But plenty of fighting and buying came afterward, and Georgia saw her share of both. Oglethorpe made his first treaty with the Indians at Savannah in 1733. It was a rather queer paper, viewed in the light of latter day transactions, and the reading of it will make smart business men wonder at the simplicity of human nature of a little more than a century and a half ago. About all that the trustees of the colony of Georgia promised the Indians in the treaty was that they would make restitution for any damage which might be done by the people of the trustees. On the other hand, the Indians agreed that the trustees' people should make use of and possess all of the lands which they had no occasion to use; and finally, to "keep the talk in their heads as long as the sun shall shine or waters run into the rivers."

Under Oglethorpe's charter from the king and treaty with the Indians, Georgia extended from the Atlantic Ocean to the Mississippi River. Since that time two other States have been carved out of the territory, notwithstanding which Georgia remains one of the largest States in the Union, ten thousand square miles greater than New York, fifteen thousand square miles greater than Pennsylvania, and only a few thousand miles less than the total area of the whole of New England. Georgia is eight thousand square miles larger than England, and has nearly half the area of the British Isles. The State contains a little more than [illegible] square miles, and about [illegible] acres. It lies between the [illegible] and [illegible] parallels of north latitude, and between the [illegible] and [illegible] parallels of west longitude.

The State Geologist divides Georgia into four geological belts, each of which has a hard name that means but little to the average lay seeker after information. Suffice it to say that, beginning at the higher altitudes, the several belts run southwesterly across the State, and by steps take the inquirer from crystalline rocks in the bold mountains to rich black alluvium on the coast. The first belt embraces the fruitful Piedmont plains, the great quarries and the mines. It is here that the golden grain nods in grateful recognition of the caresses of the breezes, and here that the hardy mountaineer by occult process converts the aforesaid golden grain into "moonshine" and "honey dew."

The mineral wealth of the section is almost beyond estimate. The hills are ribbed with the finest building marbles and granite, and girded with iron ores in quantities seemingly exhaustless. There are railroads in the State which have their roadbeds based upon marble of a quality which a prince might employ in the embellishment of his palace. Georgia marble comes in all colors and tints, from pure white to dark green. This latter, which is being quarried to a considerable extent, commands probably the highest price of any native marble. As regards granites, Georgia contains enough of them to replace that paving of "good intentions" which Dr. Samuel Johnson refers to in one of his most frequently quoted epigrams. Georgia granites and marbles are seen in many of America's great public buildings, frequently in States which themselves produce materials similar but not so good.

A list of Georgia's minerals would take in pretty near the whole catalogue, it would seem, from the reports of the State Geologist. The variety includes about everything that one could think of, and a hundred things which one could not think of without expert assistance. Gold is found in paying quantities in a dozen counties, in nuggets, in quartz and in dust. The Government at one time established a mint at Dahlonega, where [illegible] gold was coined. There has been no [illegible] that place since the War, however, improved [illegible] facilities having made it cheaper to send [illegible] gold to one of the great central mints. Silver, copper, iron, corundum, magnetite, asbestos, [illegible] soapstone, bauxite, lead and graphite are [illegible] the State. Kaolin, which [illegible] to the potteries of the North [illegible] valuable product of Georgia, and [illegible] the most refractory in the United [illegible] there are phosphate deposits, [illegible] are fertilizers by Georgia mills, or [illegible] across the ocean. Some of [illegible] building materials in the world are

mountain peaks. Between

found in unlimited quantities in this State, and in several localities [illegible] of [illegible]

[illegible]

They contain a [illegible]

employed in the m[illegible]

ing. The value [illegible]

the State will [illegible]

pine is famous a[illegible]

for ship spars [illegible]

from the earth [illegible]

Nothing is more [illegible]

with oil finish, while [illegible]

make it almost [illegible]

however, is only one [illegible]

The cypress and the [illegible]

high favor for shingles [illegible]

which never [illegible]

ment set aside [illegible]

to [illegible]

supplanted the wooden craft [illegible]

now cut by the Government [illegible]

cherry and maple [illegible]

Speaking of [illegible]

In this line G[illegible]

during the p[illegible]

[illegible]

to the seaboard. T[illegible]

[illegible]

belt begins at [illegible]

[illegible] between

[illegible] planted and growing. Still [illegible]

millionaire. It washes out political differences, and reunites in saccharine consistency friendships once estranged. So much for the poetical consideration of the melon; commercially speaking, it is one of the State's best money-makers. There is a good profit in raising melons for market, and a profit to the transportation companies in hauling them. Ten years ago the crop amounted to virtually nothing at all; now melons by the million are harvested and sold every year, and shipped to Northern markets by the train load. Fast freights take them from the fields and deliver them at the centers of population fresh, crisp and sweet. As a consequence, they usually bring a good price, and many thousands of dollars are put into the pockets of the growers. The melon belt of the State extends from the central portion in a southeasterly direction to the sea, though every county in Georgia will produce the fruit to perfection. Almost any [illegible] of vegetation common to the temperate

zone can be raised in Georgia. The State produces olives and celery, oranges and wheat, apples and chestnuts. Anything that can be grown from Florida to Washington State can be grown in Georgia, and seven times in ten better than in the majority of other places. Georgia farms are now [illegible] to cotton, corn, peas and potatoes [illegible] do not begin to trench upon the [illegible]. Any variety of [illegible] desired [illegible] found between the blue mountains and [illegible] equable, the temperature [illegible] falling. And the average value of [illegible] about $8 per acre.

As a cotton manufacturing State Georgia has made seven-league strides during the past few years. [illegible] she has been a textile manufacturer for more than fifty years [illegible] only since [illegible] that she has forged to the front as a real competitor with New England States in cotton goods. In the year named there were in the State only [illegible] spindles; now there are more than [illegible], and the investment in cotton mills approximates $7[illegible]. The greater number of the mills are operated with water power, still only a small proportion of such power available has been utilized. There are in the State a hundred falls and rapids with enormous horse-power which they offer to enterprising capital for development. The available water power in Georgia would turn the mill wheels of the United States, and leave a surplus [illegible] or Mexico. The [illegible] of the experiments in transmitting [illegible] made at Niagara [illegible] up wonderful possibilities [illegible].

Cotton [illegible]

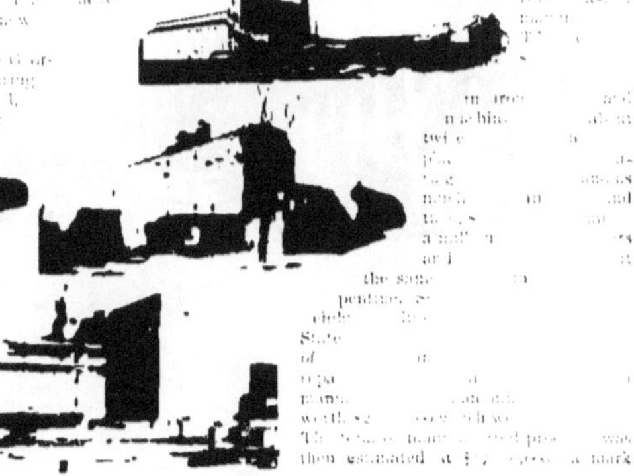

[illegible]

the same [illegible]

State [illegible]

worth [illegible]

The [illegible] was then estimated at $[illegible] a mark

[illegible] is now not less than [illegible] prosperity of the State—and that the [illegible] is evidenced by the fact that the [illegible] according to the [illegible] is due to the railroads [illegible] liberally managed, [illegible] an interest in [illegible] building up the sections [illegible]. Many of the flourishing industries [illegible] have been encouraged [illegible] aided by the railroads. The railroad [illegible] State [illegible]

[illegible]

institute, Spelman Seminary, Morris Brown College, the Baptist College, Gammon Theological Seminary, and Atlanta University, at Atlanta; Mercer University, St. Stanislaus College, Mt. De Sales Academy, and Wesleyan Female College, at Macon; the Georgia Normal and Industrial College, and Middle Georgia Agricultural College, at Milledgeville; Shorter College, Hearn Institute, and Everett Springs Seminary, at Rome; Southern Female College, at Manchester; Emory College, at Oxford; Andrew Female College, and Bethel Male College, at Cuthbert; Young L. G. Harris College, at Young Harris; Southern Female College, at La Grange; Georgia Female College, at Gainesville; Agnes Scott College, at Decatur; Levert College, at Talbotton; Clark University, at South Atlanta; and State Industrial College, at College. There are also several law and medical colleges for white students. Special schools are provided for the education of the blind and deaf. Georgia as a State has taken the highest and most advanced position in educational matters, not only in the liberality and comprehensiveness of her appropriations, but also by the adoption of modern methods of primary, intermediate and college instruction. Her public schools are recognized as models, and in her normal schools she is preparing teachers who shall be thoroughly competent to carry on the intelligent standard she has already established.

There is great liberality of opinion in Georgia. So long as an individual behaves himself and obeys the law, [illegible] the good opinion and respect of his neighbors, he is at liberty to think as he pleases, without losing [illegible] of the good will and respect of the [illegible]. Representatives of almost every Christian denomination are to be found in the State, as well as of the Jewish, the Confucian and Mohammedan religions.

The main line of the Southern Railway from Wash-

Gainesville, Flowery Bra[illegible], Ba[illegible], S[illegible]nes, N[illegible] [illegible] Caar[illegible]. The main l[illegible] ta[illegible]es due west from Atlanta toward Al[illegible] a[illegible] p[illegible] the [illegible] towns of Chatta[illegible]ochee, Mableton, Austell, L[illegible] Springs, Doug[illegible]sville, V[illegible]a R[illegible], Bre[illegible], W[illegible] and Tal[illegible]poosa.

The Chattanooga, Atlanta and B[illegible] division of the Southern enters Georgia from Tennessee at C[illegible] [illegible] [illegible] at[illegible] Atlanta, the [illegible] of [illegible] K[illegible] of Sugar Valley, [illegible] [illegible] Dallas, Powder Sp[illegible] A[illegible] Atlanta, and between [illegible] B[illegible] the road passes th[illegible] M[illegible] [illegible] cotton center, Flo[illegible] Springs, Macon, Ad[illegible] Pa[illegible] Massey, Helena, McRae, [illegible] Ha[illegible]hurst, Baxley, Satter[illegible], Jes[illegible] [illegible].

These two main lines of the Southern [illegible] N, with Atlanta as the cro[illegible] point [illegible] is the main artery of travel be[illegible] S[illegible]west and New York, and the other between the Southeast and Louisville and Cincinnati. In addition to these main lines there are numerous important branches, one from To[illegible], on the main line, to Elberton by way of [illegible]ville, another from Suwanee on main line to Lawrenceville. From Atlanta a branch runs to F[illegible] Valley by way of Williamson, where it crosses the Columbus division which runs from Atlanta via McDonough to Columbus. Upon [illegible] branch are the towns of Selina, F[illegible] Yatesville, Culloden and R[illegible] [illegible] latter, Griffin, Concord, Woodbury, Warm S[illegible] Waverly Hall and Oak Mountain. From Rome, on the Chattanooga-Atlanta division, two branches leave, one passing through Coosa and intersecting the Chattanooga-Birmingham line at Attalla, Ala., and the other meeting the Atlanta-Birmingham line at Anniston. Another division of the Southern which leaves the Washington-Atlanta line at Charlotte, N. C., enters the State at Augusta by way of Columbia, S. C.

It will thus be seen that the Southern Railway is furnishing Georgia with a most convenient and complete transportation system, with trunk line connections [illegible] the main centers north, east, south and w[illegible].

In the Greek mythology there w[illegible] a great huntress who w[illegible] with the [illegible] and as strong as he[illegible] male comp[illegible]. She was noted as a wrestler, and in a contest with Peleus threw him. She was chaste and despised love, long remaining devo[illegible] and true to Artemis, the beautiful sister of Apollo.

This huntress was Atalanta.

There was another Atalanta, told of in Boeotian legends, who was the fleetest of mortals. She was only to be won by h[illegible] w[illegible] could [illegible] her in the race, t[illegible] consequ[illegible] [illegible]. She carri[illegible] a spear, [illegible] running [illegible] Hippomenes, before sta[illegible], obtained fro[illegible] A[illegible] [illegible] three golden apples which at [illegible] in the race, he dropped, and Atalanta, stopping to p[illegible] them up, fel[illegible] behind and Hippomenes w[illegible]

former citizens many a brave man had been laid in a soldier's grave, while the women and children were scattered over the face of the earth. With peace, the [illegible] returned to build new homes. Behind these came people from neighboring States and from the North and West, and they have been coming ever since, attracted by the city's equable and healthy climate, her favorable location as a trade center, her splendid railroad facilities, the push and enthusiasm of her citizens, their unbounded faith in the future of the city, and especially by the cordial welcome extended to all.

Atlanta has a wonderful climate. In winter there is just sufficient frost and crisp [illegible] in the air to give the blood a healthy stimulus, and to nip off the bud any [illegible] blown thither. The summers are equally

free from the short, intense heat waves which smite the Northern and Western cities and the long stretches of dead heat which envelop some of her sister cities from early June until late September.

[illegible] at a map of the Southeastern States, [illegible] that Atlanta lies at the foot of the Allegheny range of mountains. The line of communication between the country on the east of that range and the vast [illegible] the west are always [illegible] rather than [illegible] difficult and infrequent mountain [illegible].

[illegible] Calhoun, after traveling the old trail to [illegible] declared with the voice of prophecy that a great city would some day rise near the [illegible] where the trail crossed the Chattahoochee River. A few years later, when the people of Georgia began to build railroads, and they were among the first in the United States to project [illegible] the first two lines were laid out

to meet at a point in the forest seven miles west of the confluence of Peachtree Creek and the Chattahoochee River. The third [illegible] State of Georgia itself, start[illegible] two were to end. [illegible] which was first applied [illegible] The name was soon after [illegible] to M[illegible] daughter of [illegible] by the way, is still living. A few [illegible] name Marthasville was [illegible] Atlanta it will remain.

The city is [illegible] elevation than any city [illegible] Rocky Mountains [illegible] pure atmosphere and [illegible] Atl[illegible] preeminence as a [illegible]

Atlanta's trade ext[illegible] M[illegible] south and beyond the [illegible] north, and from the At[illegible] to the Mississippi River and [illegible] commanding trade position [illegible] a business enterprise [illegible] center Atlanta has almost limitless [illegible] beyond the edge of Alabama's [illegible] iron fields and with nearly a dozen different varieties of iron ore in

[illegible]

the mountains of Georgia [illegible] to the north, her possibilities [illegible] scarcely be [illegible] making wooden [illegible] products [illegible] to [illegible] annually [illegible]

At[illegible]

United S[illegible] parts of [illegible]

At[illegible]

Unit[illegible] her [illegible]

Two [illegible] held in At[illegible]

[illegible]

city in close touch with all the country lying around [illegible] great distance in every direction.

By reason of cheap material building costs less in Atlanta than almost anywhere else, and imposing structures eight to eleven and twelve stories high attest the money which seeks investment in this field.

The State Capitol is here, and all about is historic ground. Just one generation ago hostile armies were battling for possession of this strategic point; but the roar of the cannon has been succeeded by the hum of spindles, the rattle of [illegible] the rattle of drays, and th[illegible] from the [illegible] smoke from [illegible] hundred manufacturing plants.

The evolution of Atlanta has been one of the wonders of [illegible] Its growth strikingly illustrates the [illegible] enterprise of its citizens, its [illegible] the wealth of the [illegible] and the [illegible] qualities of its build[illegible] Barely thirty years [illegible] began [illegible] and to properly celebrate its [illegible] Atlanta and the State produce [illegible] dream of beauty and a [illegible] of Southern [illegible]

The population [illegible] Atlanta in [illegible]

[illegible]

[illegible] Institu[illegible] the following [illegible] [illegible]tion in [illegible] Morris Brown College [illegible] Atlanta [illegible] Atlanta Bap[illegible] [illegible] There is in addition [illegible] and Agri[illegible] Scott Institute [illegible] at College Park and Decatur [illegible]

[illegible] the city, members of the Atlanta [illegible] capital and [illegible] a number of other [illegible] banking and loan companies in [illegible] the Clearing House, with [illegible] aggregate [illegible] of $[illegible]

[illegible] Atlanta has been growing, other centres of [illegible] though not at [illegible] [illegible] W[illegible] Savannah [illegible] [illegible] position [illegible] appro[illegible] [illegible] S[illegible] is the first naval [illegible] [illegible] [illegible] of [illegible] coming [illegible]

wick, a sea termin[illegible] of the Sou[illegible]ern Railway, and a [illegible] by having a wonderful amount of pluck and [illegible] [illegible] As a port Brunswick has a great future, [illegible] al present. Only recently she sent [illegible] flying the American flag. [illegible] ever shipped in one bottom. [illegible] is a railway terminus of great importance, and [illegible] both coastwise and foreign, large quantities of cotton, [illegible] lumber and phosphate rock.

The lumber trade at Brunswick [illegible] [illegible] The total volume [illegible] [illegible], coastwise and foreign, in [illegible] for the district of Brunswick for the month ending June 30, 1897, is as follows: Lumber exported [illegible] [illegible] feet; timber, [illegible] feet; cross-ties, [illegible]; shingles, [illegible]. The Board of Trade gives shipments from Brunswick alone for the year ending June 30, 1897, as follows: Lumber, [illegible] feet; timber, [illegible] feet; cross-ties, [illegible]; shingles, [illegible].

[illegible] BRUNSWICK, GA.

The New York *Times*, in a recent issue, quoted the comments of a prominent New York merchant regarding Brunswick, as follows:

"I [illegible] the trip over to Brunswick several times, [illegible] I was very much surprised at the evidences

of progress that were presented to me... Its citizens and merchants are full of activity and ... of a bright future. Business has never in the ... of the place been so profitable as it is now, and great improvements are going on.

"Brunswick's progress comes largely from its splendid location as a port from which ... manufactures of the Southern region may be shipped abroad. It is engaged in the West Indian trade and with South American countries, and is constantly handling increases in such exports. It is also becoming important as a port with direct European connections. ... escaped the attention ... that the present time ... extending ... No one could ... progress ...

... been ... lumber are ... rapidly becoming the port ...

... western parts which have been using these points as an outlet ... export ... $...

... at Brunswick ... more than ... the ... has had. The Southern Railway ... deal of the business ... improving ... trading ... there. The ... confidence in its future, for they would not ... such expense ... unless the prospect ... of making much ...

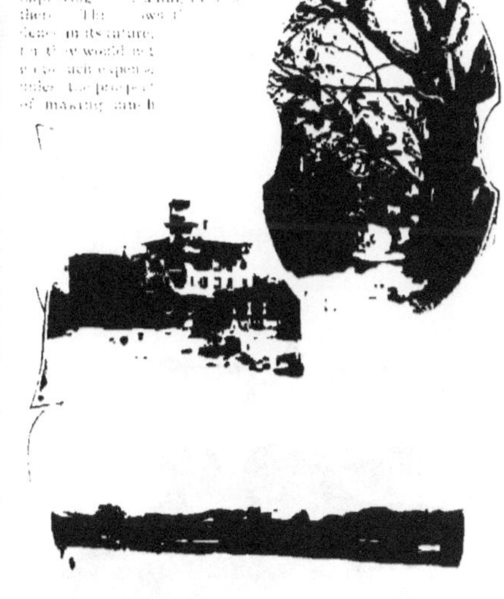

... are invested in ... and foreign ...

[illegible] coast [illegible] are many [illegible]

[illegible] white. On Cumberland [illegible] estate, "[illegible]," on which Light [illegible] Harry [illegible] General R. E. Lee, and General [illegible] Greene, Washington's most trusted [illegible] Revolutionary War. It is [illegible] has erected on it [illegible] pile of granite and [illegible] on the beach near the Government lighthouse [illegible] has been built.

The [illegible] of these islands runs back far into America's infant years, for they made a convenient [illegible] point for the adventurous explorers who for pillage or settlement cruised up and down the coast from Virginia to Florida. The first authentic mention of these islands occurs in a report made to Queen Elizabeth in [illegible] by Sir Francis Drake, who had been on a Spaniard-[illegible] down the coast, where he had [illegible] Carthagena, St. Jago [illegible] St. Augustine. As a specimen of the high perfection [illegible] attained [illegible] quotation. [illegible] and [illegible] island, [illegible] Seeing some [illegible] We un-[illegible] the [illegible]

[illegible] We further [illegible] We [illegible]

and cattle, eating what we [illegible] of the fresh beef and carrying the rest aboard our ships. Having in mind the merciful disposition of your gracious Majesty, we did not kill the women and children, but having destroyed [illegible] and their provisions and property, and [illegible] away all their weapons, we [illegible] them to starve.

"In view was another island, fifteen [illegible] concerning which we [illegible] lands if any Spaniard [illegible] thereon. The women [illegible] most ungracious, obstinate, [illegible] their husbands has [illegible]

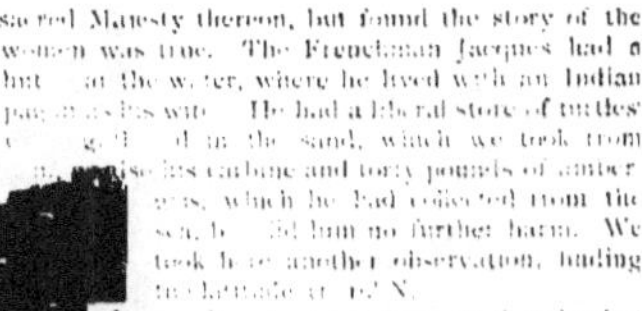

sacred Majesty thereon, but found the story of the women was true. The Frenchman Jacques had a hut [illegible] at the water, where he lived with an Indian [illegible] his wife. He had a liberal store of turtles' [illegible] in the sand, which we took from [illegible] his carbine and forty pounds of amber[illegible], which he had collected from the sea, [illegible] him no further harm. We took here another observation, finding [illegible] N.

It [illegible] assumption that the first island Drake visited was Cumberland, and the Jacques [illegible] referred to was [illegible] as Jekyl. [illegible] region [illegible] within [illegible] which gave [illegible] Bath [illegible]

killed before their eyes, and wickedly refused to [illegible] us, but after we had [illegible] a hole with a [illegible] through the tongue [illegible] most venomous of their number, they [illegible] told us that there were [illegible] Spaniards upon the other [illegible] that it was the [illegible] solitary French[illegible] Jacques, who [illegible] his own, and [illegible] it was known [illegible]

[illegible] by the devil, [illegible] important to the pat[illegible]

visitor who is strong enough to [illegible] those who are [illegible] in an invigorating atmosphere. There are [illegible] points of interest in the vicinity [illegible] pleasant [illegible] [illegible] [illegible]

[illegible] $8.50 per horse-power per annum. The source of this power is the Augusta Canal, owned by the city. This is one of the largest canals in the United States, having a capacity of [illegible] horse-power, [illegible] of which is unused and is for rent at the present time. The advantage of Augusta's low rate of $5.50 per horse-power is perceived when comparison is made with Lowell, Lawrence and Holyoke, Mass., which pay $20 per horse-power yearly, with Paterson, N. J., where the rate is [illegible], Manayunk, Pa., where it is $50.25, and Rochester, N. Y., where $25 is the rate. Lockport, N. Y., ranks next to Augusta in cheapness of water power, and there the annual rate is $11.50.

Augusta has an excellent location as a distributing center, the many wholesale houses having a large and growing business. The financial condition of the city is

distances [illegible] by [illegible] accumulation of means. The [illegible] is not so far distant from Northern [illegible] as to render the journey expensive, and living is cheap. [illegible]

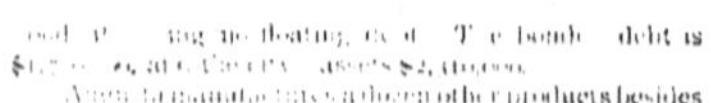

good [illegible] no floating debt. The bonded debt is $[illegible], and the city [illegible] assets $2,[illegible].

[illegible] manufactures a dozen other products besides cotton, including fertilizers, chemicals, iron and steel, and furniture. One of the most interesting sights of the city is the Confederate obelisk on the canal. This is the [illegible] chimney of the old Confederate powder mill. The site of the powder mill is now occupied by one of the finest cotton mills in the world.

Augusta is a very attractive city from many standpoints, and offers inducements of a strong character in a variety of directions. She is a clean, bright city, well built, and filled with handsome homes and charming people. Her Broad Street is a rarely beautiful and imposing thoroughfare, asphalted and well swept, and is the business artery of the city, running through it from end to end. It is traversed by the cars of a well-equipped electric system which covers the whole city.

The city is, because of its attractiveness and delightful climate, one of the most popular of all the Southern winter resorts, and is visited each season by thousands of tourists. Its location in the center of the pine ridge section of the State gives it a wonderful freedom from humidity. The beautiful, modern and handsomely fitted Hotel Bon Air is an exceedingly popular stopping place with tourists, and is one of the best-known hostelries of the South. It affords its guests every opportunity for enjoyment and recreation. There are excellent golf links at Augusta, and playing is indulged in throughout the winter.

The city of Macon, which is approaching the 50,000 mark in population, is located in the very heart of Georgia on the Ocmulgee River, which is open for navigation to the Atlantic Ocean, and is the chief center of a rich tributary territory. As a result its commerce is large and is growing steadily. The city's trade exceeds in amount over $[illegible] annually, and it has long enjoyed the distinction of being one of the leading jobbing and distributing points of the South. The region that it supplies through its wholesale trade gives in return fruit, grain, cotton, live stock, gold, granite, marble, iron, coal, lumber, turpentine and rice. The variety and abundance of raw material nearby have conspired to make Macon an important industrial center. Factories are numerous and prosperous, and $2,000,000 is invested in textile industries, as evidenced by several large cotton mills. There are also a number of foundries and ma-

chine shops, wheel works, furniture factories, cotton-seed products mills and other manufacturing establishments. They give employment to over [illegible] hands and their product reaches into the millions in value every year. Some of the largest fertilizer factories in the South are located at Macon.

Macon has one of the best public school systems in the South, and spends each year on its maintenance about $[illegible]. In higher education, too, the city occupies a commanding place. In Wesleyan Female College Macon has the first college in the world to confer academic

[illegible] was provided for by the late George I. [illegible] is doing [illegible] of higher learning, [illegible] as a city for residence, and its [illegible] has in a [illegible]

ROME, GA.

[illegible] It has [illegible] water power [illegible] the country, [illegible] street rail- [illegible] [illegible] the [illegible]

best print and cotton mills south of New England, the largest plow works in the South, and the only bagging mill in this section.

Columbus is destined to become one of the greatest industrial cities of the South. It has the advantages of a great water power, fine railroad facilities, and proximity to the coal fields of Alabama, and handles over 150,000 bales of cotton annually. Sites for factories are donated and new industries in every way encouraged, showing the enterprise of the city and the substantial invitation it extends to investors. Columbus is celebrated for its healthfulness. There has never been an epidemic in the city, a case of malaria has never been known, and the average death rate is but twelve out of a thousand. Its school system is excellent, and it was the first city in the South to establish the graded system. Modern [illegible], with all sanitary arrangements complete, are provided for pupils of every class. There are also a number of private schools, [illegible] college for women, a business college, and a [illegible] public library domiciled in its own building. The growth of Columbus has been strong and wholesome, and it is to be included in any list which may be made of the prosperous cities of the South.

[illegible]

Rome is the county seat of Floyd County, and the commercial center of one of the most attractive and progressive sections of the industrial South. It has a population of more than [illegible], and is steadily increasing in wealth, numbers and commercial importance.

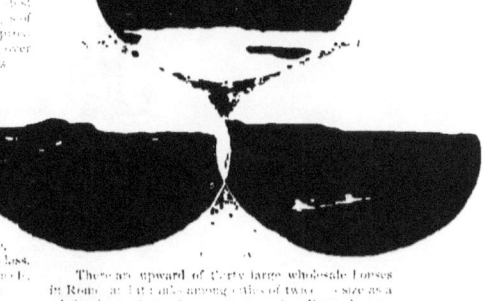

There are upward of forty large wholesale houses in Rome, [illegible] among cities of twice [illegible] size as a wholesale market. As a cotton market Rome has no [illegible] in all that section. The town has [illegible] twenty miles of well-macadamized [illegible]. Floyd County is justly noted for [illegible] good roads. There are now completed more than seventy-six miles of macadamized roads, built of hard limestone and [illegible] of easy grade and thoroughly [illegible]. There are twenty-five to thirty

schools, a seminary for young ladies, Dalton Female College.

Fort Valley is in the peach belt, the largest peach orchards in the world being located here. This is the home of the famous Elberta peach, and the center of a fine agricultural district. Fort Valley has about 2,000 people, several factories and two fruit-canning establishments. It is said that around Fort Valley there is enough hardwood timber to supply all the furniture and wagon factories in the United States for ten years.

At an altitude of ... feet above sea level, and within

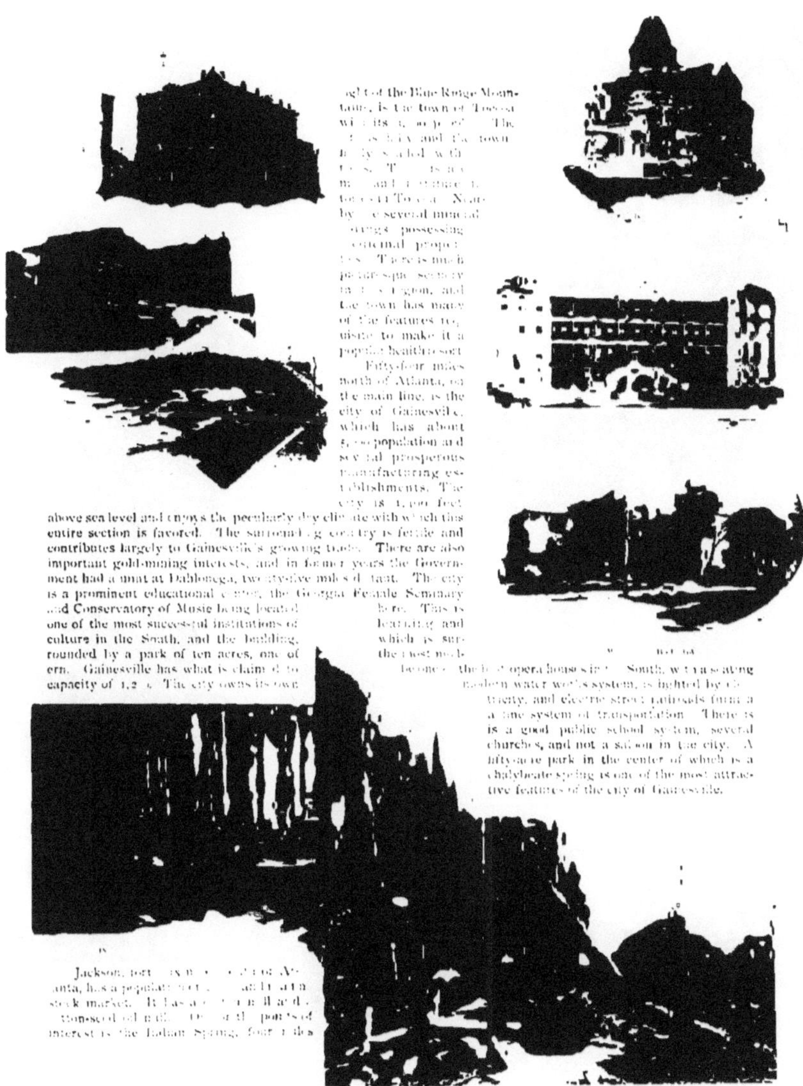

[illegible] of the Blue Ridge Mountains, is the town of Toccoa [illegible] its [illegible] The [illegible] and the town [illegible] several mineral springs possessing [illegible] properties. There is much picturesque scenery in this region, and the town has many of the features [illegible] to make it a popular health resort.

Fifty-four miles north of Atlanta, on the main line, is the city of Gainesville, which has about [illegible] population and several prosperous manufacturing establishments. The city is [illegible] feet above sea level and enjoys the peculiarly dry climate with which this entire section is favored. The surrounding country is fertile and contributes largely to Gainesville's growing trade. There are also important gold-mining interests, and in former years the Government had a mint at Dahlonega, twenty-five miles distant. The city is a prominent educational center, the Georgia Female Seminary and Conservatory of Music being located here. This is one of the most successful institutions of learning and culture in the South, and the building, which is surrounded by a park of ten acres, one of the most modern. Gainesville has what is claimed to be one of the [illegible] opera houses in the South, with a seating capacity of 1,2[illegible]. The city owns its own modern water works system, is lighted by electricity, and electric street railroads form a fine system of transportation. There is a good public school system, several churches, and not a saloon in the city. A fifty-acre park in the center of which is a chalybeate spring is one of the most attractive features of the city of Gainesville.

Jackson, [illegible] Atlanta, has a population [illegible] stock market. It has a [illegible] points of interest is the Indian Spring, four miles east of the town. There are several churches, and the educational establishments include the Jackson Institute with over 300 students.

Tallap[illegible]sa is a town of about 2,[illegible] population. The elevation is nearly [illegible] feet above sea level, the climate good. The town has modern improvements, several factories, three hotels and seven churches. Two miles from town a gold-mining company is operating with about one hundred men. Considerable attention has been given to grape culture recently, more than [illegible] acres being planted in grapes nearby. Tallapoosa Lithia Springs Hotel is an excellent and popular resort.

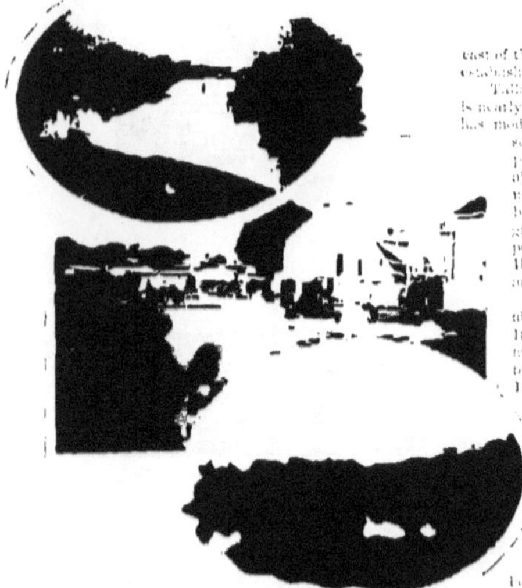

Elberton contains about [illegible] people. It has a cotton mill, five churches and, by way of schools, Elberton Collegiate Institute, Elberton Seminary and Bowman Institute, a colored school. It is a thriving town, and has considerable neighboring trade.

Eastman, on the line between Atlanta and Brunswick, is one of the [illegible] young cities of G[illegible] [illegible] of [illegible] enterprising citizens is rapidly [illegible]

[illegible] which has recently set in toward Georgia is a [illegible] both the State and the newcomer. Homes [illegible] at the hands of a generous and warm-[illegible] to the newcomers. Evidence [illegible] settled by colonists [illegible] State [illegible] in every county in the [illegible] officers of the Federal Army, [illegible] [illegible] land [illegible] and [illegible] have [illegible] in the State, and a number

[illegible] of attractive resorts, both [illegible] Brunswick, St. Simon's Island, and Cumberland Island, on the coast; Mt. Airy, in the northeastern part of the State; Lithia Springs, twenty miles west of Atlanta; Tallulah Falls, Indian Springs, near Flovilla; Tallapoosa Lithia Springs, Warm Springs, New Holland Springs, and a number of others of less prominence.

Lithia Springs are twenty miles west of Atlanta, and are famous for the curative properties of the water, which is stronger in lithia than that from any other spring in the country. There

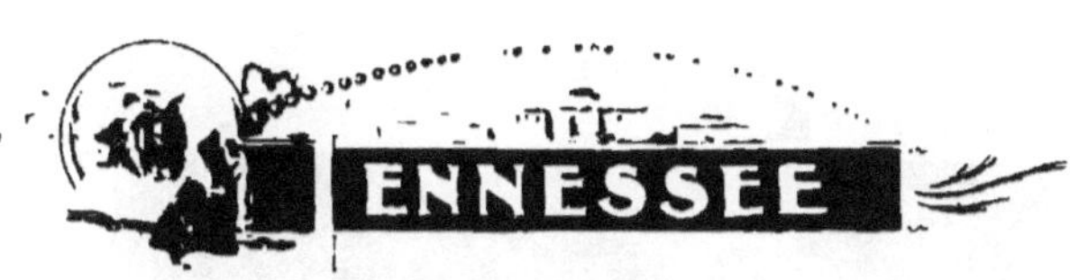

TENNESSEE has [illegible] started on the second [illegible] of her statehood with a swinging stride that betokens her confidence in her destiny and [illegible] strength to fulfill it. From a period of [illegible] she advanced [illegible] of promise with all the hopefulness of conscious power.

With [illegible] the hospitable gates [illegible] splendid capital [illegible] past period [illegible] before all the world the fruits of progress [illegible] the hundred years that are gone, marking the span between a [illegible] primeval [illegible] and an industrial empire, and forming a stupendous object lesson of man's triumph over Nature. Her Centennial Exposition was [illegible] of the world into her confidence. She [illegible] of the sleeping [illegible] of coal and iron in her bosom, of the forests on her mountains, [illegible] power dashing [illegible] down her rivers, [illegible] the [illegible] under cultivation [illegible] fields [illegible] diversity [illegible]. Throughout [illegible] she has been prodigal of opportunity, and her children [illegible] that prodigality [illegible]. The [illegible] has been [illegible] State with over [illegible] its development has been [illegible] examination [illegible] and [illegible] sufficient [illegible] for inducing immigration and [illegible].

The State [illegible] from that given by the Indians [illegible] River [illegible]. As early as [illegible] North Carolina [illegible] mountains and located in the [illegible] western [illegible]. In 1776, when [illegible] constitution of North Carolina was [illegible] settlements [illegible], the District of W[illegible] sent [illegible] representatives to the legislative body, [illegible]. The [illegible] State neglected [illegible] the mountain [illegible] ceded [illegible] to the United States. The [illegible] formed the State of Frank[illegible] and [illegible] Franklin [illegible] John Sevier [illegible] Governor, [illegible] by Congress [illegible]. In [illegible] these [illegible]

divided naturally into three grand [illegible] the eastern, middle, and western. The last-named lies [illegible] the Tennessee [illegible] Mississippi rivers [illegible] of [illegible] is the [illegible] State, which [illegible] year [illegible] It [illegible] be mentioned as a [illegible] distinction in [illegible] of Tennessee that she stands among the great cotton manufacturing States as [illegible] in the United States. There are in the State [illegible] cotton mills, [illegible] spindles, 2,314 looms, and [illegible] bales were consumed [illegible] last year.

[illegible] comprising forty-[illegible] counties, [illegible] Tennessee River [illegible] the Cumberland [illegible] and scattered over it are some

of [illegible] grain and fruit farms in the world.

[illegible] Railway [illegible] through eastern Tennessee [illegible] resources is [illegible] from Bristol to [illegible] southeast to [illegible] to Birmingham, Ala., [illegible] to Memphis.

[illegible] through the entire [illegible] Southern Railway runs, is two [illegible] average width of [illegible]

The [illegible] very [illegible] choicest for farming and dairy purposes [illegible] who are tired [illegible] is made [illegible] north-west [illegible] ores, [illegible] where streams [illegible] east of this iron [illegible] valley to its south [illegible] and ledges of marbles, [illegible] lead and other mineral [illegible] it is again bordered [illegible] ores, famous for the [illegible] made from them. The [illegible] fine timber, clear streams [illegible] and beautiful scenery [illegible]

[illegible] Bean, Holston, Clinch, Pigeon, [illegible] Emory, [illegible] and Watauga Rivers [illegible]

The valley is dotted with [illegible] intervening valleys of great fer[tility] [illegible] with success corn, wheat, [illegible] the various grasses. Nearly all [illegible] in the

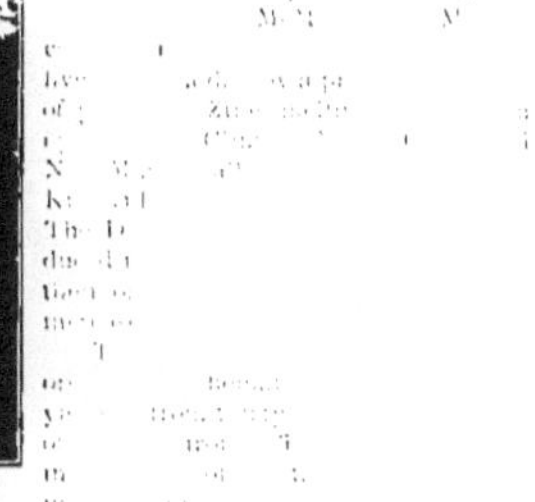

region, and hundreds of thousands of dollars' worth of fruits and berries, green and dried, [illegible] canned, are annually shipped. Apples, [illegible] pears, plums, strawberries and grapes are the principal fruits. Vegetables grow to surprising perfection, and some of the finest market gardens [illegible] are adjacent to Knoxville.

The climate is equable [illegible] healthfulness. The growing season [illegible] about one hundred [illegible] The winters are mild, and the summers for the most part delightful. The average altitude is one thousand feet above sea level. [illegible] are no extremes of [illegible] tornadoes, the mountains [illegible]

[illegible] on the Virginia line [illegible] Hot Springs on the North Carolina border [illegible] Chattanooga in the lower end of the valley [illegible] valleys abound in rivers and streams of pure water, clear [illegible]

in marketable quantities [illegible] taken out in open [illegible] being done. There [illegible] Cranberry [illegible] Johnson City, [illegible] Dayton [illegible]

Tennessee [illegible] on the market [illegible] hundred thousand tons of iron ore [illegible]

Britain did in 18[illegible], more than the United States did in 1842, and half as much as was produced in the United States in 1861. Though considerable advance had been made in this industry prior to the War, the period of most rapid development dates from 1873, when investors first began to realize the wonderful opportunities presented for the employment of capital and energy in connection with the great natural advantages. In the production of both red and brown hematite ore, Tennessee occupies [illegible] place among the States of the Union. Magnetic and other varieties of ore are found in paying quantities, but not to such an extent as the above named.

An illustration of the remarkably superior methods of iron making to-day over those of ante-bellum times is shown by the statement that with forty-nine furnaces, in those earlier days, Tennessee produced [illegible] tons, at a cost of $[illegible] per ton, while any one of the larger furnaces in the State to-day would produce more iron than all the forty-nine did in 18[illegible]. Large numbers of men are employed in the industry, and it is one of the great sources of wealth to the State.

The coal fields within a radius of sixty miles of Knoxville embrace about one hundred thousand acres, including the Jellico and Coal Creek districts from which about one million tons are mined annually. The annual production of the State is about [illegible] tons, wholly of the bituminous variety. The coal field of Tennessee covers [illegible] miles, [illegible] in the east and eastern middle sections of the State.

Probably no section in America produces more or a greater variety of [illegible] clays. It is estimated that there is a [illegible] already named are thrown out among the *débris* at the coal regions. These clays usually [illegible] from fifty-five to sixty-five per cent of silica, and from twenty to [illegible] per cent of aluminum. [illegible] abundant. [illegible]

general industries. When to these is added transportation facilities and unlimited

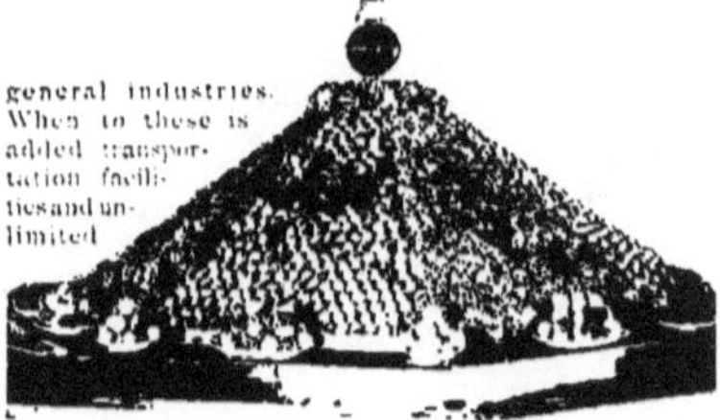

AN EXHIBIT OF TENNESSEE FRUIT

water power, everything is present for the most prosperous development.

East Tennessee is noted for its marble, which exists in practically inexhaustible quantities and almost endless variety and tints. There are now about one hundred quarries in operation, which produce 300,000 cubic feet a year. Forty of these are in the vicinity of Knoxville, which is one of the largest marble shipping points in the United States. The marble columns and balustrades in the Capitol at Washington are from Hawkins County, east Tennessee, and some of the finest quality is seen in the new Congressional Library Building. New York's Capitol at Albany is finished in Tennessee marble, and so are a number of the handsomest buildings in Chicago, including the Public Library, New York and other cities.

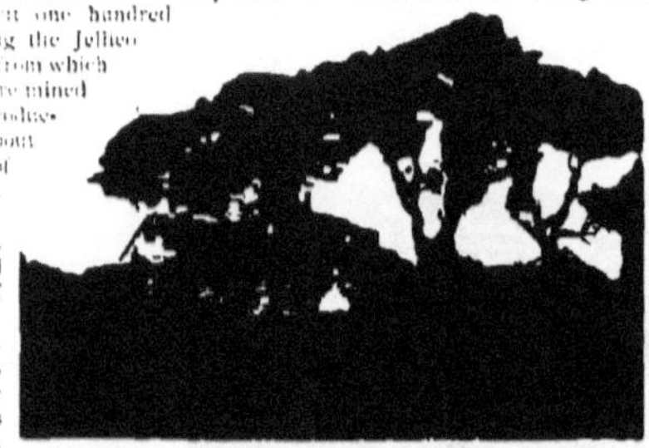

A TENNESSEE HOMESTEAD

A large amount of timber has been shipped from east Tennessee, but the supply is practically inexhaustible. Hard woods of the finest quality abound. Oak and poplar are most abundant, but hickory, cherry, chestnut, walnut, maple, beech, sycamore, ash, persimmon, dogwood, basswood, sourwood, sassafras, gum, hemlock, buckeye, spruce and balsam are among the most plentiful woods. There are [illegible] acres of forest lands in the State, and the lumber output from [illegible] saw mills located in every division of the State is over [illegible] feet annually.

The growing of tobacco is one of the most important industries in Tennessee, and in some sections as a money crop it takes the lead. There are eighteen States in the Union denominated as tobacco growing States. Of these Kentucky leads in number of acres and value of the crop, North Carolina next, then Virginia, and Tennessee fourth. The number of acres planted in Tennessee in [illegible], number of pounds raised, [illegible], [illegible]

With [illegible] and scenery so picturesque it is most natural that east Tennessee should [illegible] health resorts. There are numerous mineral [illegible] resorts where

FALLS OUT FROM [illegible]

thousands spend the [illegible]

[illegible] Normal School, East Tennessee Female Institute, [illegible] College, [illegible], Maryville College at [illegible], Tennessee College, [illegible] Sweetwater Military Academy, [illegible] University and [illegible] College, [illegible] Ward Seminary, [illegible] Normal [illegible] Nashville [illegible], Jackson, [illegible] Southwestern [illegible] Clarksville Female [illegible], Columbia, [illegible] Cumberland University, [illegible] Hughes School [illegible] White College, Fayetteville [illegible].

[illegible]

States of the Union. She is equally [illegible] education, having over [illegible] public [illegible] valued at $[illegible] the curriculum [illegible] and high schools [illegible] the State [illegible] learning. [illegible] in her State [illegible] being the [illegible] the South, [illegible]

The University of Tennessee [illegible] Knoxville [illegible] very justly [illegible] system of the State [illegible] the [illegible] of 18[illegible] ritory now [illegible] attention is [illegible] tific [illegible] sical [illegible] from the [illegible] there are [illegible] either the [illegible] established [illegible] cation [illegible]

industry a million dollars is invested and [illegible] persons [illegible] employment.

The city's advantages in all lines of manufactures, in the way of cheap [illegible] proximity to raw materials, excellent transportation, and connections with central markets, are [illegible] recently receiving the attention they have merited. As a consequence it is now confidently expected that the textile industry will be brought to its rightful place of coordinate importance with the iron and wood [illegible].

While Chattanooga is preeminently a manufacturing city, its [illegible] interests are important and growing. As a jobbing center it has the great advantage of being the natural commercial [illegible] [illegible] [illegible] re[illegible] for [illegible] two hundred miles [illegible] which [illegible] [illegible] well. Radiating from the [illegible] throughout this [illegible] lines of railroad, and flowing by is [illegible] river transportation to [illegible] controlled by [illegible] Alabama Great Southern Railroad completely encircles the city, and [illegible] are located some of the largest [illegible] in this [illegible].

[illegible] Chattanooga [illegible] not allow [illegible] of prosp[illegible] factories to [illegible] the importance of good [illegible]. It points with [illegible] to the [illegible] that last year [illegible] per

cent [illegible] its [illegible] between [illegible] world
one were [illegible]
the [illegible]
Grant [illegible]
schools [illegible] Park
and [illegible]
In the [illegible]
[illegible]
Americ[illegible]
proving every [illegible]
material advantag[illegible]
out forgetting the things
at the highe[illegible]

No pen has [illegible] ever will adequately p[illegible]

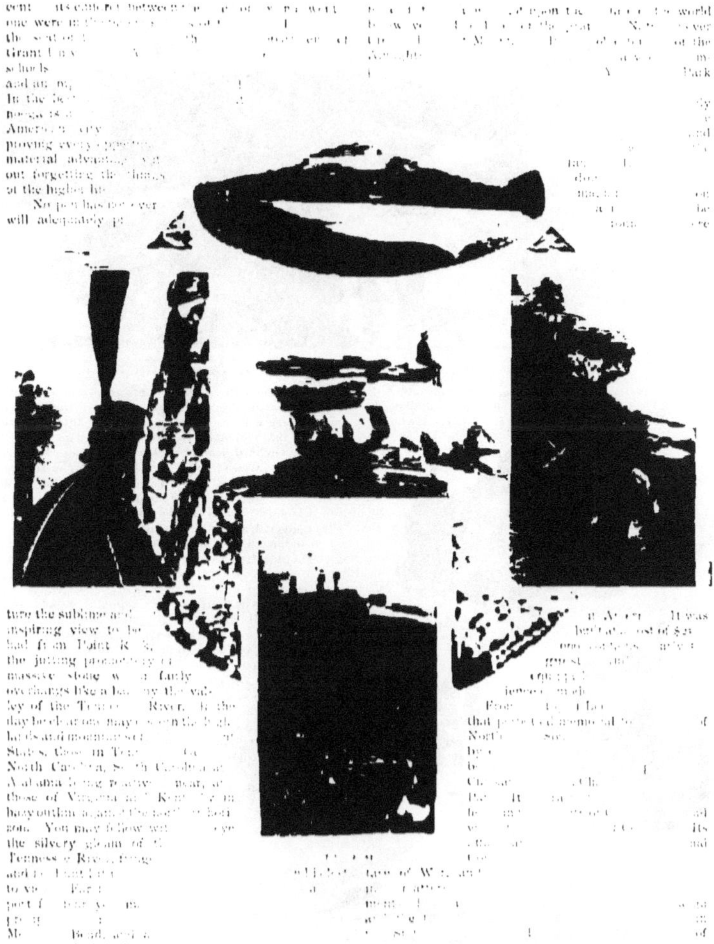

ture the sublime and [illegible] inspiring view to be had from Point R[illegible], the jutting promontory [illegible] massive stone w[illegible] fairly overhangs like a ba[illegible] the valley of the Tenn[illegible] River. [illegible] the day be clear one may [illegible] the highlands and mountain[illegible] States, those in Ten[illegible] North Carolina, South Carolina [illegible] Alabama [illegible] those of Virginia and Kent[illegible] hazy outline against the [illegible] zon. You may follow wi[illegible] eye the silvery gleam of [illegible] Tennessee River [illegible] and [illegible] to [illegible] [illegible] M[illegible] Bend, [illegible] tan[illegible] which is [illegible]

[illegible] It was [illegible] at a cost of $[illegible] [illegible] From [illegible] that [illegible] memorial to [illegible] of North [illegible] by [illegible] Ch[illegible] [illegible] its [illegible]

[illegible] tare of W[illegible]

The battlefield on Lookout Mountain is a portion of the Chattanooga division of the park. Practically, the city of Chattanooga itself is also a portion of this division, since by State and county laws and city ordinances the Park Commission is given authority to mark all points of military interest with tablets and monuments, and the jurisdiction of the Government over the same has been assured by law.

THE MOCCASIN BEND OF THE TENNESSEE RIVER
FROM LOOKOUT MOUNTAIN, CHATTANOOGA IN THE DISTANCE

All of the roads used by troops in the battle have been restored and improved by the best methods of road-making known to modern engineering. The mileage of driveways throughout the park amounts to about sixty miles.

The Chickamauga field consists of about five thousand acres of woodland, all of which has been cleared of underbrush and through every part of which a team can drive without difficulty, and about fifteen hundred acres of open field. The brigade lines of battle upon seven distinct fields, namely, Chickamauga, Brown's Ferry, Wauhatchie, Orchard Knob, Lookout Mountain, Missionary Ridge, and Ringgold, have been accurately identified with the assistance of twenty-six State Commissions interested, and a large number of participants in the battles, a[illegible] them. Most of these lines are already marked by monuments, and by historical tablets. About one thousand historical tablets have already been erected, and a large number of locality and distance tablets and other guides to movements upon the fields. All fighting positions of batteries for both sides on the Chickamauga field have been indicated by the erection of guns of the same pattern as those used by the battery in the engagement, upon iron gun carriages which are an exact reproduction of those of the battle. Thirty-five battery positions on one side and thirty-three on the other have been thus marked by the mounting of over two hundred guns. A majority of the battery positions thus far ascertained in the Chattanooga section of the park have been marked in the same way.

The underlying element of the park establishment is the restoration of the battlefield. By the clearing out of timber which has grown since the war, the closing of new roads, and the opening of the roads of the battle, the Chickamauga battlefield has been restored in almost every respect to its condition at the time of the battle.

Both the Northern and Southern States which had troops engaged are actively at work in ascertaining the regimental lines of battle of their troops and marking them by monuments.

The tablets which are erected by the National Commission are strictly historical. These tablets show the organization of armies, corps, divisions and brigades with their respective commanders, the brigade tablets carrying these designations to the commanders of regiments and batteries. The historical text upon each tablet giving from 250 to 500 words, is very carefully prepared and then passes through the hands of each member of the National Commission and their historians, and finally must receive the approval of the Secretary of War before being erected upon the field. The same method is observed in regard to all inscriptions upon monuments. The locations of all monuments, markers and tablets must also receive first the approval of the National Commission, of which Gen. H. V. Boynton is Secretary, and finally that of the Secretary of War, before they can be erected.

One of Tennessee's most progressive cities is Knoxville, which was founded in 1792. It was the first capital of the State, and the original capitol, which is still in a good state of preservation, is an object of interest to all visitors to the city. Among Knoxville's distinguished residents have been Gens. John Sevier, the hero of King's Mountain, Andrew Johnson, Davy Crockett and many others.

to be built. Knoxville has more bridges than any other city of its size in the country. Two magnificent structures which span the Tennessee are occupied by railroads, and two are for the use of the public, one of which, a new bridge costing nearly a quarter of a million, is about completed. Several other bridges span the creeks which border the city upon two sides.

The river is navigable for boats for seventy miles above Knoxville, and below to the Ohio. The city ranks fourth in volume of trade among the cities of the South. It has many wholesale houses, a large proportion of them doing an exclusive jobbing business. The volume of trade has been estimated at about $40,000,000 annually. A careful canvass of the jobbing houses shows that they employ 1,300 people, and their trade extends to every one of the Southern States.

Knoxville's ten banks have a total capital of $1,1[illegible] and a surplus of $150,000. Bank clearings for 1897 amounted to $21,4[illegible] [illegible] clearings [illegible] $[illegible]

The University of Tennessee, now in its second century of its existence, maintains in every department [illegible] and honor. No educational [illegible] The University is [illegible] system [illegible] State [illegible] may [illegible] and deaf [illegible] their appropriate [illegible]. The university buildings, [illegible] are beautifully located on an eminence in West Knoxville. The university structures tower above the city like the battlements of some ancient castle. The campus is [illegible] with beautiful oaks, among which [illegible] endless archways of walks and drives, bordered by grass plots, shrubbery and flowers. Here "classic shades and leafy dells" are a charming reality.

Among the other institutions of learning at Knoxville are the Morris Classical School and the East Tennessee Female Institute, Fountain City Normal School, Knoxville Medical College, Knoxville College (colored), Baker and Himmel University Preparatory School, and the State institution for the deaf and dumb.

Knoxville's residence section is particularly attractive and demonstrates in the numerous beautiful homes the refinement and culture of its inhabitants. Few if any of our American cities of equal population have more to be proud of in the line of municipal improvements. The social atmosphere of the place is of the best, [illegible] population is made up of sturdy, enterprising people who are ever alert to the city's welfare, and [illegible] that makes for progress, [illegible]

Bristol, [illegible] Tennessee and Virginia,

[illegible] occupies a unique position, the boundary line between the two States running along the center of Main Street. While preeminently a commercial city, its institutions of learning are among the finest in the South, including three colleges (one male and two female) and

fields of ... ... North Carolina, in the ... west of T...

Alleghany Mountains ...

Johnson C...

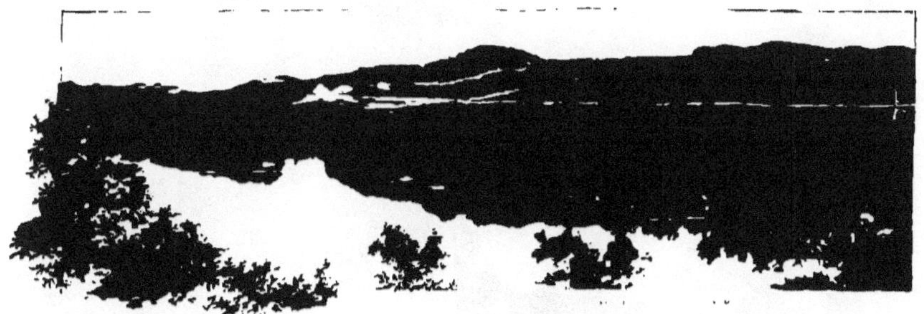

compared with the other cities in the State. Situated among the foothills of the Great Smoky Mountains on the south, and the beautiful peaks of the Clinch on the north, with the valleys of the Nolachucky and Holston rivers lying between, it has an excellent natural position [illegible] of unparalleled beauty and grandeur.

A [illegible] and manufacturing point, Greeneville [illegible] of a circle whose radius of one hundred miles [illegible] fields of coal, iron, marble, granite, slate and virgin forests of oaks and other timbers suitable for manufacturing purposes.

On either side of the Southern Railway are rich [illegible] whose products are [illegible] as the seasons themselves. On

the slopes of the Great Smokies are orchards in which the fruits never fail, while the quality compares with those of more Southern climes. Between these and the Nolachucky are the bright tobacco fields where [illegible] golden [illegible] has made [illegible] for wrappers and taken the first [illegible] at Cincinnati, New Orleans, Richmond, Nashville, and other points of exhibit. Three large tobacco factories and warehouses are located in Greeneville, which [illegible] large quantities, but export [illegible] to Europe [illegible] have prospered [illegible] from Maine to Mexico. [illegible] the other side of [illegible] and are given to [illegible] as well as [illegible] a great [illegible] Greeneville during the [illegible] for poultry and [illegible] North [illegible] markets.

[illegible] of President [illegible] his residence [illegible] the monu[illegible] by almost every [illegible] the Southern Railway.

[illegible] town. Ten miles north of Morristown, and at the south[illegible] base of Clinch Mountain, are the [illegible] Springs, a most attractive health resort and one of great popularity with [illegible] of the Southern States who [illegible] its attractions. [illegible] guests spend [illegible] here and find [illegible]

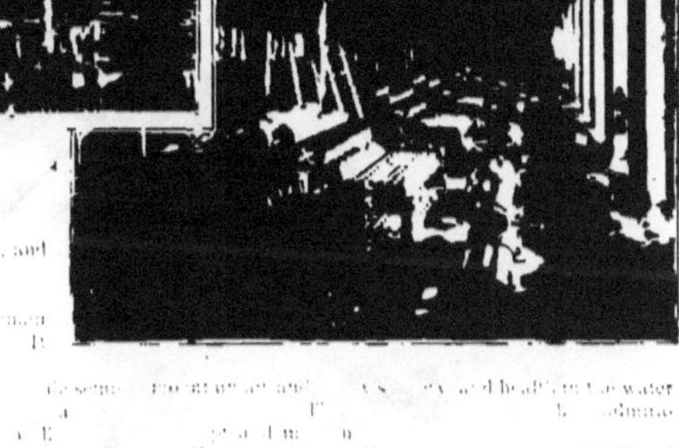

Among educational institutions may be mentioned Greeneville and its [illegible] College, coeducational, founded 1794; numerous public and private [illegible], and a colored college.

Rogersville is situated [illegible] Southern Railway, [illegible] line, between Greeneville and M[illegible]. It is a progressive [illegible] educational [illegible]. The [illegible] nary, has [illegible]. Other [illegible] colored students, [illegible] Rogersville is Hale Springs, [illegible] popularity as a health resort [illegible] good hotel.

[illegible] scenic mountain and [illegible] and health in the water [illegible] Southern Railway has [illegible] a large [illegible] central [illegible]

Morristown [illegible] east of Knoxville [illegible] Southern Railway [illegible] It is the county seat of Hamblen [illegible] the famous counties [illegible] Tennessee. [illegible] central location in east Tennessee makes it a good trading point with ample [illegible] facilities. Morristown has two [illegible] two tobacco factories, woodworking and other [illegible]. There are [illegible] churches, and a [illegible] building costing $2[illegible] besides a normal academy. A very desirable fruit growing and general farming district surrounds the

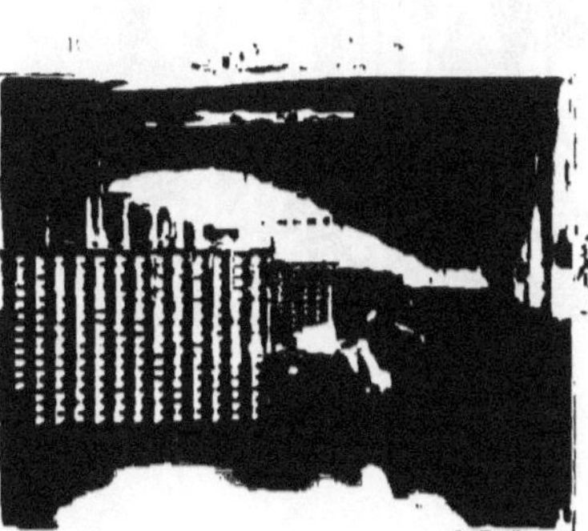

[illegible] of Knoxville [illegible]

[illegible] from [illegible] Chattanooga [illegible] stand [illegible] the [illegible]

INTERIOR [illegible]

[illegible] of Athens [illegible] with [illegible] Its [illegible] East Tennessee. Among its industries are two cotton mills, one woolen mill, several flouring mills and others. [illegible] Athens is [illegible] U. S. [illegible] University maintained by the Methodist Church [illegible]

[illegible] Athens [illegible] where the Southern Railway crosses the picturesque Hiwassee River [illegible] modern [illegible]

of the most prosperous small towns in the South. About [illegible] miners are now employed in [illegible] and almost [illegible] daily, the most of it [illegible]. The coal [illegible]

about it are some of the best producing coal mines in [illegible] Tennessee.

[illegible]

[illegible]

nually paid out [illegible]. These coal deposits are inexhaustible, as the mountains for miles are [illegible] with the "black diamonds." [illegible] abounds in the [illegible] mountains, poplar, pine, [illegible] white oak, hick[illegible] many other valuable hard [illegible] woods. The country around Jellico is well adapt[illegible] to the growing of [illegible] and [illegible] garden

[illegible] sale or manufacture of intoxicating [illegible] company.

The [illegible] was [illegible] most remarkable ever known. [illegible] for as much [illegible] days after the [illegible] plots were [illegible]. Within a year's time [illegible] Harriman [illegible] and is keeping [illegible] of prosperity. It has many [illegible] water works, electric light, well paved and sewered streets and [illegible]

[illegible] Harriman [illegible] twelve [illegible] of [illegible] the [illegible] Springs, a noted health and pleasure resort, situated on the [illegible]

[illegible] There [illegible]

[illegible] The population of Jellico [illegible]

Another important coal town is Coal C[illegible] Anderson County [illegible]

From Chattanooga the [illegible] the [illegible] Railway extends to Memphis, which is exactly [illegible] miles [illegible] [illegible] leaves [illegible]

[illegible]

Tennessee, and is not only important commercially, but exceedingly interesting to archæologists, who have made it the center of extensive investigations into the prehistoric mounds which are [illegible] found throughout this immediate section. The [illegible] Wolf and Hatchie rivers wind through the [illegible] county, merging

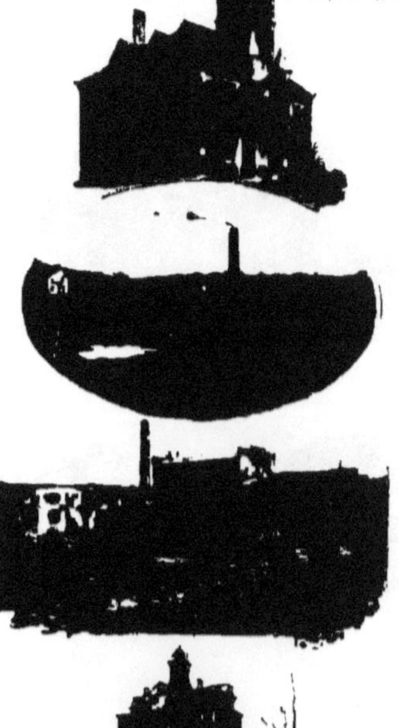

just before their confluence with the Father of Wat...
and upon the summit ... one of ... Indian ...
junction is an immense ...
attention of students ... both ...
tries. The consensus of opinion ...
more years ... ago the ...
County ... were till ...

evidences of a forgotten people. In 1541 the intrepid
De Soto and his ...

was from the ...

annual average
bale valued

the southern

year, and

and almost
street.
Memphis is
public library
house, a

cotton and

ALABAMA

[illegible] the Government census it was only [illegible]

[illegible] first of the States of the Union [illegible] by [illegible] the [illegible] Bigbee, [illegible] power [illegible] both [illegible] of [illegible] advantages [illegible] many [illegible] great staple, [illegible] growing to [illegible] on the continent [illegible] It is a fact that an Alabama cow [illegible] world's butter record, [illegible] running over a year.

Flourishing orchards and nurseries tell their own story to the eye. The new and growing colony of Fruithurst speaks for the [illegible] country [illegible] of [illegible] two years, already grown [illegible] of an [illegible] The [illegible] and [illegible] have a wide reputation, while market [illegible] space [illegible] Anniston and [illegible]

But [illegible] all, the chief [illegible] advantage [illegible] mountain [illegible] the pros[illegible] to the [illegible] deposits, [illegible] rapidly [illegible] which [illegible] for all [illegible]

[illegible] Alabama [illegible] land [illegible] This is [illegible]

poplar, ash,
hickory and gum
all of which a
being cut in
quantities that
make impor-
tant contribu-
tions to com-
merce and th
wealth of the State.
But by far th

The
for
hand,
To the wealth of

most important are

a few small mines. In 1897 she produced [illegible] tons from [illegible]

In [illegible] the production of [illegible] was [illegible] tons. In 1897 it was [illegible] tons, made from [illegible]

In [illegible] the ore mined was [illegible]. In 1897 [illegible]

In [illegible] the production of [illegible] was [illegible] tons. In [illegible] it was [illegible]

In [illegible] the market was local and uncertain and hard to [illegible]. Now Alabama coal supplies the [illegible] consumption in nearly a half dozen States on the South Atlantic and [illegible] Gulf, goes by large contracts to the railroads of Mexico, runs engines in the ports of foreign countries, and through the Southern Railway's [illegible] lines of rail and barge is driving Penn- [illegible] [illegible] of the great Mississippi [illegible] divided [illegible] into three [illegible] them, the [illegible] and the Warrior. The last named [illegible] square [illegible] The Cahaba field

metropolis of the country. Not [illegible] with meeting every demand as to quality and underselling every competitor in all the markets of America, [illegible] Alabama makers have invaded Europe, and from April [illegible] November [illegible] exported to England and [illegible] Continent over 200,000 tons of iron. [illegible] furnace in Alabama is [illegible] than anywhere else in the world.

This is not all [illegible] the story. Alabama [illegible] seller of the raw material, cotton, coal and iron in the [illegible] Her iron pipe and foundry industries have grown to immense proportions and she is an active and earnest competitor in all the markets of the world for the sale of such products. Her iron foundries send engines and boilers and sundry articles [illegible] the South [illegible] chinery for the [illegible] sugar [illegible] and [illegible] for them [illegible] out every portion [illegible] world.

Alabama [illegible] ment forward [illegible] facture of [illegible] the best open [illegible] steel at a price [illegible] scale that has [illegible] stand [illegible] and [illegible] market [illegible] steel raw pig [illegible] [illegible] the State [illegible]

that it ushers in, will not fail to impress and modify the [illegible] of the people in the profoundest and most far-reaching way, [illegible] the change already begun whereby the population is being turned from the simpler methods of agriculture to the more varied and intense and enterprising activities of manufacture. The most important step to the front Alabama has ever made was accomplished during the summer of the past year by the Birmingham Rolling Mill Co., which put into operation its first furnace, with a capacity of sixty tons a day, and scored a success from the initial run. Another furnace of the same capacity is already built and in operation, the [illegible] making a mere beginning of the steel era now [illegible] inaugurated.

Perhaps the best idea [illegible] mining and [illegible] industry has [illegible] in Alabama [illegible]

[illegible] in which [illegible] has done [illegible] development in the South [illegible] In one respect the opening [illegible] of Alabama [illegible] part of the State [illegible] and [illegible]

[illegible]

City, near Gadsden, the experience being such that a company of Boston mill men are following their example and are now [illegible] at Cordova, [illegible] Walker County coal [illegible] new [illegible] at Montgomery.

[illegible] industrial activity [illegible] the State. Industries built as [illegible] the specula[illegible] and [illegible] business enterprises. This is [illegible] at Sheffield and Decatur and Anniston. There is effort everywhere, the spirit [illegible] and the spirit is the prelude to [illegible]

The agricultural [illegible] mining and manufacturing development of Alabama within twenty years presupposes several conditions about which the investor as well as the emigrant is wont, and very properly, to inquire closely. Progress of this sort must be accompanied by steadily increasing transportation facilities. Accordingly we find in Alabama [illegible] miles of the best-equipped railroad lines, the track of the Southern Railway alone in the State being [illegible] miles. Four important

[illegible] and are [illegible] Every

Industrial success is accom[illegible], too, with [illegible] institutions which make [illegible] complete—well [illegible] revolent societies, [illegible] for moral and intelle[illegible] the State [illegible] and towns almost as [illegible] sity at Tuscaloosa [illegible] equipped; and [illegible] Mechanical Col[illegible] stitute at Au[illegible] white girls at Mont[illegible] an agricultural school [illegible] district; five normal sch[illegible] for negroes, in all of w[illegible] agriculture and the mechanical arts are a large [illegible] of the course, all doing [illegible] work. Nothing could better illustrate how [illegible] abreast Alabama is with the times than the fine equipment and successful [illegible] of her school for the deaf, dumb and [illegible], where these unfortunates are taught at the State's expense all that modern science and art permits them to know. The public [illegible] of the cities are as good as can [illegible] cities of the country wit[illegible] Those of [illegible]mingham, Mobile and Montgomery especially Birming[illegible] school

Atlanta Exposition [illegible]. All political parties in Ala[illegible] the importance of education, and planks [illegible] maintenance of the [illegible] are embod[illegible] Besides what the State does for the [illegible] and the private [illegible] at work. In all [illegible] State [illegible] the [illegible]

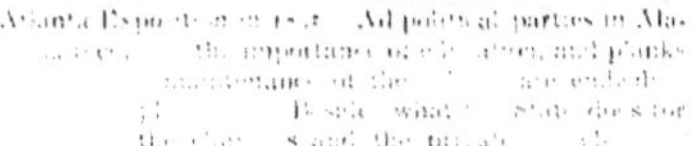

[illegible]

shown by the co-education of the sexes, Alabama affording to her girls every facility that she gives to her boys. No reference to education in Alabama would be complete without reference to the important work being accomplished by President Booker T. Washington at the Tuskegee Normal and Industrial Institute, of which he is the moving and controlling spirit. This institution may be termed the most successful of any in the country designed for the education of the colored [illegible] organized by President Washington [illegible] and a leader [illegible] property valued at $[illegible] which [illegible] acres of land, [illegible] handicraft and labor [illegible] buildings. It has [illegible] which [illegible] every [illegible] men and young women [illegible] industrial training [illegible] [illegible]

and continues to Meridian, Miss., 37 . miles southwest ...

... passes through Fort Payne, and intersects ... at Attalla. Between this point ... Meridian it passes through ... York and ... intersecting the main east and ...

... Southern, with the same ... and Meridian, Miss. ...

... Montevallo ...

trunk line and is both for passenger and freight traffic
to the . . .

. . . industry . . .

prospects and . . .

Jefferson County . . .

population . . .

old, . . .

be the . . .

centers . . .

Birmingham . . .

increased . . .

iron furnace in . . .

today there are . . .

among the . . .

Sloss, . . .

in the . . .

from which are . . .

coal daily. Birmingham . . .

trial hive, and . . .

been the . . .

. . . to the . . . laborers and mechanics. In Birmingham alone there are . . . not less than twenty- . . . industries . . .

The [illegible] of Birmingham have a capital of $1,[illegible] [illegible] carry deposits aggregating $[illegible], with annual [illegible] of about $[illegible]. The aggregate

[illegible] for at the Atlanta [illegible] The schools [illegible] an average [illegible] the public schools [illegible] Highlands Academy, [illegible] Birmingham Medical College, [illegible] Art School, [illegible] and two business [illegible] Presbyterian [illegible] western [illegible] several [illegible] water [illegible]

electricity [illegible] steam. Six railroads center here. The altitude of Birmingham, which varies from [illegible] to [illegible] feet above sea level, and the remarkably fine surface drainage, added to the excellent sanitation by modern sewers, makes the city one of the healthiest in the South, the death rate being but [illegible].

Anniston is one of the flourishing new cities of the South. Its manufactures comprise car works, foundries, [illegible] factory, saw mills and almost every variety of plant [illegible] in a city of its size, including one of the most extensive establishments in the country for the casting of car wheels.

The chief resources upon which the city was founded and [illegible] are coal, iron, lime, wood and cotton. It is located in the center of the brown hematite iron ore district of Alabama, and this iron has become celebrated through its adaptability for the manufacture of car wheels and axles, and these have stood a higher test than those made in any other part of the United States. Building material is cheap, and buildings can be constructed for one-third less than the same would cost in New England.

The healthfulness of Anniston is based upon its having an altitude of [illegible] feet above the sea level; its being in a mountainous region, removed from any local or nearby influences calculated to produce ill health; furnished with a bountiful supply of the purest spring water, which is distributed throughout the city through an excellent modern water system; a complete modern system of sewerage, which conveys the sewage of the city into a large stream four miles distant, and with the

[illegible] to perfectly drain [illegible] within [illegible] very short time [illegible]

The city has a well-equipped fire department, which [illegible] and prompt in the performance of duty, and an excellent electric system, which lights the city and supplies power for the street railways. Anniston is surrounded by a country most fertile and productive [illegible] valley farms [illegible] supply [illegible] vegetables, [illegible] poultry and [illegible] The city has a [illegible]

[illegible]

of Alabama known as the "Black Belt." It is the capital of Dallas County, which is in itself larger in area than the State of R[illegible] Island, and is located on the top of a high bluff [illegible] looking the noble Alabama

[illegible]

[illegible]

In the suburbs of Tuscaloosa are the State Insane Asylum and the State University. The campus of the latter institution is perhaps the most beautiful in the South. It comprises about forty acres of land, in the form of a square, and is almost as level as a table. Set well to the rear are four large buildings so arranged as to constitute the university quadrangle. The main avenue leads south from the front of the quadrangle, directly toward the president's mansion, which is across University Avenue, a beautiful drive from the city of Tuscaloosa.

Established in the early thirties, the history of the university is in large measure the history of the State. Many of the leading citizens of the State are its alumni. Nor is its influence confined to Alabama. The institution has furnished many distinguished men to other States in all walks of [illegible]. The land of the university site is about [illegible] acres, the same being valued at $[illegible]. The buildings, of which there are seven, are valued at $[illegible],000, while the libraries, cabinets, apparatus, etc., are valued [illegible], a grand total of $[illegible]. The university owns [illegible] acres of the best coal lands in the State, worth at present prices about $[illegible], but these lands are rapidly increasing in value, so that the productive value within the next few years will be at least $[illegible].

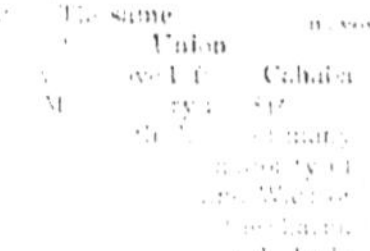

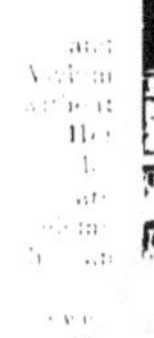

There are two general departments of instruction, viz., academic department and department of professional education. The academic department embraces four courses, leading to as many degrees, which are, bachelors in mining, and in civil engineering, science and arts. In the professional line, the university has a law department in Tuscaloosa and a medical department in Mobile. Taken as a whole the university is one of the best-equipped institutions in the entire South.

Fruithurst is situated on the line of the Southern Railway, in Cleburne County, in the northeastern portion of the State, [illegible] miles west of Atlanta, Ga., and [illegible] miles east of Birmingham, Ala.

In the spring of 1895 the city was started by the Alabama Fruit Growing & Winery Association, who purchased 20,000 acres of fruit lands and located in the center the city of Fruithurst. There has been literally hewed out of the woods at this point in two years and a half an incorporated city of 500 people, with 150 houses, stores, a hotel, free school,

and the Fruithurst Inn, costing $10,000. Upward of 2,000 acres have been planted to grape vines, over 5,000 acres sold, and a total of nearly $600,000 in actual cash expended in improvements.

The station of the Southern Railway is located at the foot of Central Avenue, with the Fruithurst Inn at the head of the avenue, a quarter of a mile away. Surrounding the station, fifty acres are devoted to experimental gardens, orchards and vineyards, in which every variety of fruit is planted.

The grape-growing industry of Fruithurst has made the town a very prosperous one.

Large quantities of table grapes will from this time on be shipped from Fruithurst to Northern markets, and probably from 50,000 to 100,000 gallons of wine manufactured.

The plan of the Fruithurst Company is to a large extent co-operative, it selling its lands in [illegible] tracts with two acres of each tract planted, caring for the vineyards of such purchasers as do not locate, and buying the crop when the vineyards are in bearing.

This is pre-eminently a grape-growing section, 5,000 acres being already planted to grapes within a radius of ten miles of Fruithurst.

The elevation being from 1,000 to 1,200 feet above sea level, the location is remarkably healthful, with pure, rarified atmosphere, freestone water, perfect natural drainage, and from a climatic standpoint it cannot be excelled.

Talladega, the county seat of Talladega County, is located in the heart of the county, two miles from the stream bearing its name, and is in every respect worthy to be the center of the county so richly endowed by Nature. The name is an Indian corruption of the Spanish "Terre Denga," or borderland, and the beautiful spring, gushing forth its hundred gallons per minute, and located in the heart of the town, has for a hundred years been the meeting place for the red races of the west and the [illegible] of the east. Historically, it is famous in many respects, and especially so as being the famous site of the battle of Talladega, fought between Jackson's forces and the Indians during the Creek wars. At present it is a modern city of six thousand energetic people, its architecture is substantial and rich, both in its business blocks and its beautiful homes. Railroads, car shops, cotton mills, foundries, all give it the hum of industry so musical to the industrious ear, while extensive business houses, equipped with all modern conveniences, make its world of commerce reliable and successful. Agriculturally, industrially, financially, morally and intellectually she is the pride of her people.

The city has many churches and a model public school system, extensive car shops, a successful cotton mill, modern water works, utilizing the waters of the great spring, and many industrial establishments, including immense iron works and a finely equipped coke furnace.

She has [illegible] about her [illegible] with improved [illegible] enough to supply a place of [illegible] manufactures. She has as [illegible] and as healthy [illegible] there are [illegible] State.

There are located here the State Blind Asylum, the Talladega Military Academy, the State Deaf and Dumb Asylum, the State Asylum for [illegible] Blind, [illegible] for the [illegible] the State Asylum for the [illegible] Deaf and [illegible] Presbyterian Ladies' [illegible]

town, and from Lookout Mountain [illegible] one s[illegible] and Sand Mountain on the other, coal and several superior iron ores are mined in abundance, making this section second in mining in the State to the Birmingham district. The population of the town is conservatively estimated at 1,500. Among its industries are an iron furnace, a large foundry and machine shop, and an extensive iron ore mining plant, one cotton gin, one cotton mill, and three distilleries. There is a graded system of free public schools, of large attendance, and four churches. An electric light plant is owned and operated by the city. The fine system of water works has been established at a cost of [illegible].

[illegible]

There is a back [illegible] of thirty miles, in which everything grows except some tropical fruits. Cotton, corn, wheat and oats are the staple products, and from the Sand Mountain district come potatoes, peaches, apples, grapes and watermelons of the greatest perfection.

Gadsden is the county seat of Etowah County, in northeastern Alabama, ninety [illegible] miles south of Chattanooga and fifty-two miles west of Rome, Ga. It is located in the fertile Coosa River Valley, and its [illegible] ments are picturesquely beautiful and attractive. [illegible] in the heart of the mineral belt of Alabama, [illegible] vast ore deposits almost at the city's limits. The city has a large trade in [illegible] handling [illegible] bales a year, and its cotton mill operates [illegible] spindles, 1,000 looms, employing [illegible] hands. There are other extensive manufacturing interests at Gadsden, and much enterprise among [illegible] citizens. The Coosa River, which is navigable for 150 miles, has been greatly improved by the Government, and [illegible] have been built at Gadsden. The city has a number of handsome [illegible] and many evidences of thrift. There are many churches and excellent schools, including [illegible] Mount [illegible] Lookout Institute [illegible].

Piedmont, in [illegible] County, is [illegible] way between [illegible] and [illegible] [illegible] from Chattanooga [illegible] Birmingham, Ala. [illegible] where [illegible] and Nances Creek [illegible]

known in the State as the Coosa Valley. This valley, from five to thirty miles wide, is a continuation of the Tennessee and Virginia valleys which lie between the ridges of the great Appalachian chain of mountains, which extends parallel with the seacoast northeast into New England, and it is in no respect less fertile, picturesque and attractive than the most productive of the family and succession of valleys of which it forms so important a part. At this Cross Plains (which was the original and most descriptive name of the town) is a great area of gently undulating lands, lying between [illegible] and [illegible] feet above the sea level, through which flow many bold streams, on which there are extensive forests of varied woods, and numerous open farms, and around which are [illegible] mountains whose altitude above tidewater varies from 1,200 to 2,200 feet.

The Coosa Valley was the last of their possessions east of the Mississippi that was surrendered by the semi-civilized Creeks and Cherokees in [illegible], and it was quickly occupied by intelligent and thrifty farmers from North Carolina, Tennessee and Georgia, [illegible] of some means, who brought with them, in addition to horses, kine, swine and agricultural implements, pleasure carriages, pianos, paintings and books. In [illegible] Major J. K. Dailey came from North Carolina and erected [illegible] and store on the site of the [illegible] town. Cross Plains, [illegible] was then called, grew slowly, and in [illegible] that it was [illegible] town. In [illegible] and [illegible] company was [illegible] developments [illegible] with [illegible] new [illegible] most [illegible]

[illegible]

county [illegible] County, an [illegible] numerous [illegible] It [illegible] the [illegible] years [illegible]

[illegible] have [illegible] years, and [illegible] Western farming [illegible] than twenty-five [illegible]

costing over $[illegible]. The city is situated on an undulating table land, [illegible] feet above sea level, which stretches from the State line on the north to the great bend of the noble Tennessee

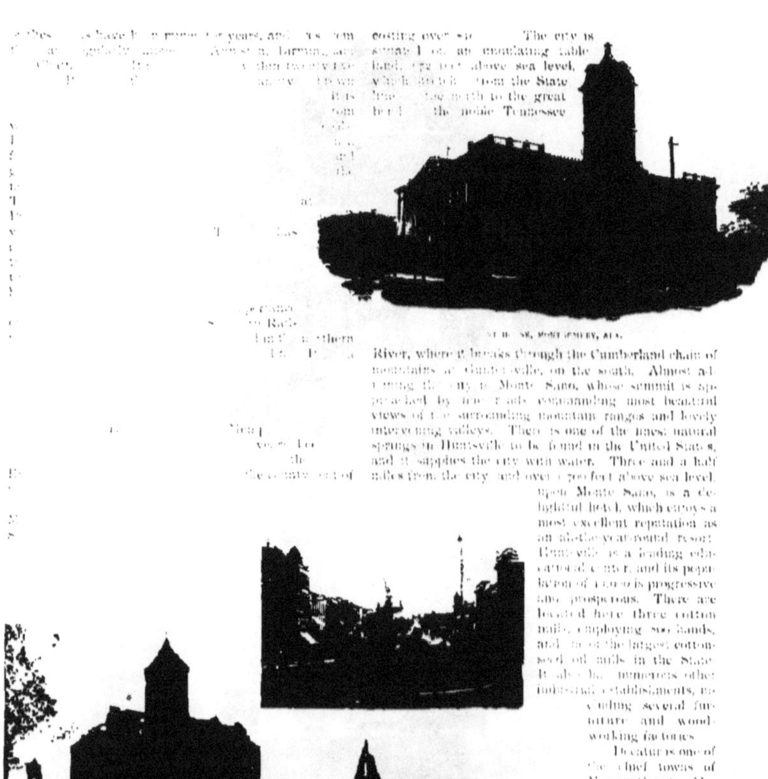

[illegible] HOUSE, MONTGOMERY, ALA.

River, where it breaks through the Cumberland chain of mountains at Guntersville, on the south. Almost adjoining the city is Monte Sano, whose summit is approached by fine roads commanding most beautiful views of the surrounding mountain ranges and lovely intervening valleys. There is one of the finest natural springs in Huntsville to be found in the United States, and it supplies the city with water. Three and a half miles from the city and over [illegible] feet above sea level, upon Monte Sano, is a delightful hotel, which enjoys a most excellent reputation as an all-the-year-round resort. Huntsville is a leading educational center, and its population of [illegible] is progressive and prosperous. There are located here three cotton mills, employing [illegible] hands, and one of the largest cotton-seed oil mills in the State. It also has numerous other industrial establishments, including several furniture and wood-working factories.

Decatur is one of the chief towns of Morgan County, Alabama, which is so fertile and so well adapted to agricultural pursuits that it has earned the soubriquet of the "Clover County." The Memphis division of the Southern Railway [illegible] the Tennessee River at this point [illegible] a handsome iron [illegible]

1887 had 1,200 inhab[illegible] and to-day has [illegible], and has progressed with wonderful rapidity in every featur[illegible] It is one of [illegible] to [illegible] pride as [illegible] of the South [illegible] from [illegible]

equally important, the cit[illegible] is a vantage of them. Among th[illegible] of Decatur are the Oak Extract works and [illegible] paving good $20[illegible] and [illegible] building [illegible] The city has a [illegible] system [illegible] plant and street railways [illegible] residences and several excellent ho[illegible] There are [illegible] churches and several well [illegible]

The [illegible] rare beauty [illegible] River, the sixth larg[illegible] in the United St[illegible] purpose by bringing the products of the valley from both directions to the markets at Decatur.

Sheffield has had [illegible] any town in Ameri[illegible] Its [illegible] when its site was [illegible] enterprising and far-seeing [illegible] The city is beautifully [illegible] stretching back [illegible] overlooks the Tennessee River [illegible] It [illegible] business [illegible] and [illegible] any city of its [illegible] have a [illegible] in [illegible] to [illegible]

[illegible] county is good, and Sheffield and Tuscumbia, its neighbor, are provided with educational advantages [illegible] are exceptional. Almost all of the religious [illegible] active organizations in the county. [illegible] Tennessee [illegible] the county [illegible] with the Memphis [illegible] of [illegible] Railway [illegible] road [illegible] in length, which pa[illegible] through Sheffield. It [illegible] and [illegible] One of [illegible] manufacturing [illegible] importance the industrial establishments [illegible] an extensive wagon works [illegible] shoe factory [illegible] The [illegible] has a [illegible] water and gas works and a paid fire department. [illegible] across the Tennessee [illegible] bridge span the [illegible] is [illegible]

The [illegible] the center of a large trading district [illegible] It [illegible] and [illegible]

In [illegible] River [illegible] Montgomery, [illegible] of Alabama.

[illegible] will be attracted [illegible] portions [illegible] progressive, healthy city. The wide streets, the splendid system of [illegible] and the [illegible] natural drainage all combine in making it one of the healthiest cities in the

There are [illegible] kinds of lands around Montgomery—the black [illegible] particularly suited to the growth [illegible] and cotton; the red lands, suited to the raising [illegible] vegeta[illegible] and fruits, and the pine lands which are [illegible] cheapest lands of the section.

and fair laws. The State, county and municipal tax amounts to two and one-eighth on a three-fourths valuation of property, and there are no vexatious laws. Montgomery has seven lines of railway. In addition, the Alabama River is an artery of travel to Mobile. The city has a population of over [illegible], an assessed valuation of [illegible] in the county, two electric car systems, is abundantly supplied with deep artesian water, a fine system of electric lights, and has fifty churches of every denomination and creed, both white and colored, a splendid system of public schools, beautiful shaded streets, and a people who will cordially welcome the visitor and the home-seeker.

The city of Mobile is situated at the mouth of the Mobile River, just at the point where it empties into the beautiful land-locked bay of the same name. It is Alabama's only seaport, and its location at the outlet of a river system aggregating more than a thousand miles of inland navigation gives it a commercial position of the greatest value and importance. Mobile ranks second only to New Orleans among the Gulf ports.

Mobile has long been the natural trading center for a large expanse of domestic territory, and can now lay claim to a rapidly expanding trade with Mexico, Central and South America, and the West India Islands. It has a population of about [illegible]. The city is laid out regularly, and most of the [illegible] streets are luxuriously shaded. The city sits upon a rolling sandy plain, backed on the west by high hills, filled with springs, from which Mobile's splendid water supply is taken.

Mobile city and county have one of the best public school systems in the South. There are twenty-seven schools within the city limits, including the Barton Academy, where most of the young people finish whatever [illegible] they [illegible] in school. The schools are [illegible] equally, both in [illegible] and style, for white and colored [illegible]. The private schools are numerous. The M[illegible] of Alabama is a State institution. It [illegible] a large attendance from all parts of the South. The parochial schools of the Catholic church are well managed and successful. The Academy of the Visitation [illegible] of extensive reputation. The same [illegible] at Spring Hill the flourishing college of St. Joseph, under the management of the J[illegible], and is now [illegible] by private endowment [illegible] school for boys in the central part of [illegible] Mobile [illegible] are of the best. It is the [illegible] of the Episcopal Bishop of Alabama and of the C[illegible] Bishop of Mobile. The Catholic Cathedral is [illegible] the forty churches [illegible] the city. [illegible] Mobile will find excellent drives in the [illegible] city, [illegible] in almost every direction. [illegible] attractive is the Bay Shell Ro[illegible] along the western edge of Mobile B[illegible] is shelled and [illegible] con[illegible] magnolia, bay, [illegible] to the [illegible] moss [illegible] spring; [illegible] Most of the [illegible] Mobile [illegible] or

[illegible]

KENTUCKY

FROM the days, over a century ago, when the hunters [illegible] started the timid deer feeding by her rivers, up to [illegible] ent, Kentucky has been a veritable [illegible] of [illegible] lapsing years [illegible] change in the kind [illegible] Instead of the roving game meat in plenty to the pioneers, there are now herds [illegible] and the [illegible] the world has [illegible] have been supplanted [illegible] planted, [illegible] rich harvest and the [illegible] Kentucky [illegible] may [illegible] to the [illegible]

The [illegible] early settlers came over the mountains from [illegible] track of their tired [illegible] eyes of [illegible] forest [illegible] to the [illegible] of the [illegible]

Their settlement [illegible] complete [illegible] together [illegible] Kentucky [illegible] waiting [illegible] Kentucky the [illegible] of man. This human element [illegible] and [illegible] came with the first settlers [illegible] the world [illegible] and a wilderness [illegible] changing [illegible]

[illegible] through [illegible] 1798 [illegible] State, bearing [illegible] own [illegible] and [illegible] and [illegible] stock [illegible]

[illegible]

States. The quality is so high that it fixes the standard in many kinds. Her Burley tobacco is so much superior to that grown elsewhere that the State has a virtual monopoly of this staple. As indicating the profits of tobacco culture, crops frequently average from 1,200 to 1,[illegible] pounds to the acre, and the choicest leaf frequently sells at $[illegible] per 100 pounds. Besides the Burley the dark or heavy types are largely grown. These do not command so high a price as the former, but as their yield is larger their production proves quite as profitable to the grower.

Kentucky leads the Union also in the production of whiskey, having just completed her first century in its manufacture. The beginning of the industry was the direct result of the whiskey war in Pennsylvania in 1798. At its close there was an exodus of distillers over the mountains to the wilds of Kentucky. The first distillery was built in what is now Mason County, but when the State was organized it was a part of Bourbon County, hence the name given the product to distinguish it from the eastern brand, which was distilled from rye.

It is interesting to note, in this period of general currency discussion, that in the early days whiskey was the principal medium of exchange in Kentucky. It possessed the currency requisites of improving with age, of ready divisibility and of portability, to say nothing of the fact that it was in great demand.

As indicating to what dimensions the industry has grown, the General Government receives in revenue alone

about $25,000,000 annually [illegible] the Kentucky product. Measured in money it is the leading industry in the State, and pays out every year millions of dollars for corn, rye and malt, in addition to the enormous amounts for labor. On May 1, [illegible] there were in the distilleries and bonded warehouses of the State 70,509,121 gallons of whiskey. For the fiscal year ending June 30, [illegible], the production of bourbon and rye whiskey was 17,[illegible] gallons.

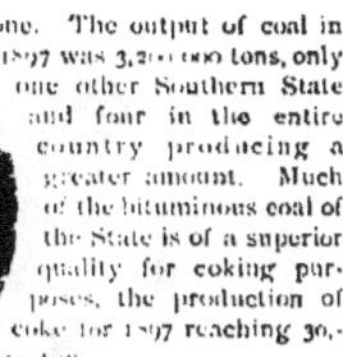

ISVILLE,

The thoroughbred is the third in the triumvirate of products in which Kentucky leads the Union. The Kentucky horse is invincible. It is claimed that there is some subtle quality in the climate of Kentucky and some peculiar nutriment in the grasses raised in her limestone pastures that unite in producing the perfect horse. Be that as it may, the fact is fully established that the blooded horses of the "Blue Grass State" excel those raised elsewhere in speed, endurance and beauty.

Mules are also raised in large numbers, especially for supplying the cotton and sugar districts of the Southern States, and to call a mule a Kentucky mule has always added several dollars to its value.

Another branch of the live stock industry which is highly profitable is feeding cattle for the European trade. These cattle, called "export cattle," are as fine beeves as are produced anywhere in the world.

Sheep raising is also followed profitably and every year more widely, particularly among the smaller farmers.

Fruit growing has claimed much attention in recent years. At the World's Fair in Chicago, Kentucky peaches were awarded the first place on account of their delicious flavor and rich coloring. In several counties peach raising has been entered upon extensively and the profitable results have become an incentive to peach culture in many other parts of the State.

When the pioneers entered Kentucky, with their axes they literally hewed out a commonwealth. The fertile farms which their sturdy work cleared lessened, of course, the area of the forest lands, but Kentucky to-day is by no means a treeless plain. Indeed, the State is well timbered, and every species of tree known to her latitude is found in abundance. The raw material is thus offered for a large wood manufacturing industry. At present the value of timber floated to market in rafts and shipped by rail represents many millions of dollars annually.

The State's mineral resources include coal, iron, zinc, vast beds of onyx, clays for the manufacture of all grades of pottery, and valuable quarries of sandstone and limestone. The output of coal in 1897 was 3,2[illegible] tons, only one other Southern State and four in the entire country producing a greater amount. Much of the bituminous coal of the State is of a superior quality for coking purposes, the production of coke for 1897 reaching 30,[illegible] tons.

In manufactures the State is on the threshold of a splendid period of development. Much has already been done in many lines, but when the possibilities are considered it seems only a beginning. With coal to feed the fires of her factories, with her hills yielding the best qualities of iron ores, with forests growing timber for every variety of wood-working, with her production of one-quarter of the world's supply of tobacco, and with a splendid system of waterways and railways for transporting the abundant raw materials to factories and their finished product to market, Kentucky possesses every essential to industrial greatness.

But this greatness is by no means only in prospect, much of it, in fact, in the manufacturing of furniture and agricultural implements, a large industry, is already thriving. In the manufacture of tobacco and cigars the State is fast coming to the position to which her primacy in the production of raw material entitles her. As has already been stated, she now leads in the production of whiskey. There are in the State several large cotton and woolen mills whose success clearly indicates what may be expected in the development of this line of manufactures.

But Kentucky has not allowed herself to become engrossed with the creation of wealth to the exclusion of the cultivation of the mind. She is generous in her provision for schools. In 1897 she was providing instruction for 736,[illegible] of her children, for which she was paying over $3,[illegible]. On teachers' salaries alone the sum reached the large total of $2,[illegible]. At the head of the school system is her State University at Lexington, with which the graded schools are federated. There are also many other institutions of higher learning, among them being Georgetown College, Central University, Center College and the State Agricultural and Mechanical College.

The cities of Kentucky are among the most progressive in the South. They are centers of enterprising activity and have always been a strong factor in the development of the State's resources. Their citizens have been quick to perceive Kentucky's natural advantages and ready to take the lead in improving them. Kentucky's cities have also become famous as centers of a charming hospitality and of all the graces of social life.

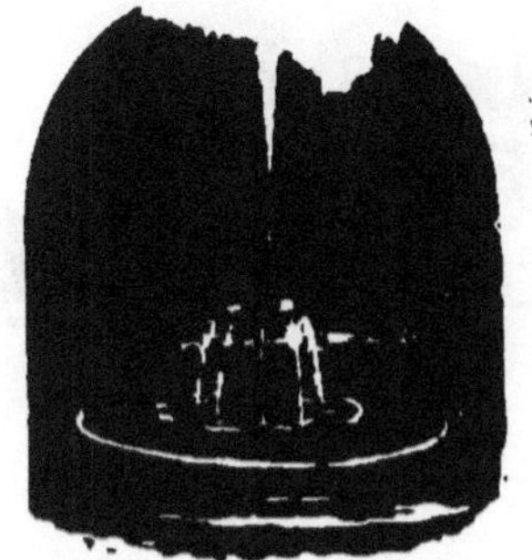

Just n[illegible]th of Lexington is the growing city of G[illegible]rgetown, with a p[illegible]tion of [illegible]. The Georgetown Bap[illegible] C[illegible]ge [illegible] located h[illegible], as well as a Catholic [illegible]. The public sch[illegible] are excellent. G[illegible]getown [illegible] a large trade [illegible] the fertile a[illegible] c[illegible] about it. It is a notable [illegible]

Near the [illegible] Middlesborough. Found[illegible] in [illegible], by an E[illegible] syndicate [illegible] its growth was rapid. [illegible] the famous Cu[illegible] and [illegible] [illegible]

[illegible]

the gateways of the Southern Railway system, will be appropriate at this time.

The fi[illegible]t settlement of Cincinnati, by Israel Ludlow, with about [illegible] other persons, occurred December 28, [illegible]. The settlement was called "Losantiville" up to January, [illegible], when it was given its present name, in [illegible] of the Cincinnati Society of Officers of the Revolutionary War. The corporate limits of the city at present comprise [illegible] square miles, and the population is something over [illegible]. On the Kentucky side of the Ohio River, immediately opposite Cincinnati, are the cities of Covington, with a population of 37,371, Newport, having [illegible], and Bellevue, Dayton, West Covington, Ludlow, and other towns, with street-car and railroad commutation connections, aggregating a closely estimated population of [illegible]. This gives a population south of the river of [illegible], which should not be excluded from any estimate of Cincinnati's population.

The business of Cincinnati is very varied, with numerous manufacturing interests, wholesale houses of all kinds, and a large jobbing trade. The traveling salesmen of its business houses may be found throughout the entire United States. It is not only the commercial center of the State of Ohio, but, being closely adjacent to Kentucky and the South, has an enormous trade throughout that section also.

The Chamber of Commerce, at the southwest corner of Fourth and Vine streets, is one of the most strikingly handsome buildings in the city. Its magnificent proportions, great [illegible] beauty, and remarkable strength and solidity are apparent to even a casual observer. The entire cost of this building was about [illegible].

Among the city's notable structures is the United States Government Building, completed in [illegible]. It contains the Post Office, Custom House and Federal Courts of the United States Government, and offices for the various departments of the internal revenue, [illegible] service, railway mail service, etc.

The University of Cincinnati, which occupies commodious and well-appointed buildings, has a very large patronage, and is endowed to the extent of nearly a million dollars.

Cincinnati's Public Library occupies its own home, and owns [illegible] books and [illegible] pamphlets, which are constantly being added to.

[illegible] many commercial [illegible] surrounding the city. The street railways [illegible] part of the city the many [illegible] which Cincinnati is justly proud. [illegible] Mt. A[illegible] [illegible] College Hill, [illegible] and Hyde Park.

[illegible] charming parks among [illegible] 20[illegible] acres in which is [illegible] and Art Academy. [illegible] beautifully improved wood [illegible] of the Cincinnati [illegible] one of the largest [illegible] [illegible] acres. [illegible] Washington Park, a [illegible] of the city, and [illegible] Park, [illegible] equestrian statue [illegible] Gen. [illegible] and the handsome statue of [illegible]

[illegible] Zoological Gardens, [illegible] any amusement [illegible] acres, and the [illegible] permanent structures [illegible]

IT is not possible to better express a general idea of the resources of Mississippi than in the language of Mr. Carlisle "This noble commonwealth is essentially and pre-eminently an agricultural State. Nature designed and fashioned it to bless and reward the labors of the husbandman Its geological formations appear to exclude it from the profits of the mine and quarry, but what the State lacks in mineral resources, sometimes transitory and always in the end exhaustive, is more than counterbalanced by a generous, responsive soil, an almost ideal climate, and productions the value of which is not excelled in any part of the Union. The first Europeans who trod its soil the adventurous and romantic expedition of Hernando de Soto—found its surface richly carpeted with the native grasses, and maize or Indian corn, one of the chief foods of mankind, 'of such luxuriant growth as to produce three or four ears to the stalk.' No State in the Union has been more liberally endowed by Nature with all the conditions favorable to agriculture. In one sense of the word Mississippi is still a new State, with its immense natural advantages as yet mainly unappropriated. Its great forests of valuable woods have been comparatively little depleted; many of its numerous fine mill and manufacturing sites await the power of skill and capital; more than half its area remains untouched by the husbandman, while the part already in cultivation may be made to double its productive power by improved methods of agriculture."

But despite the fact that general geological appearances seem to be against it, there are many who hold firmly to the belief that portions of the State contain extensive coal beds. There are distinct traces of coal along the edge of the hills bordering the Yazoo Valley on the east, especially in Holmes County near Tchula, where tests on an extensive scale are contemplated But, as has been well said, Mississippi can waive all pretention to mineral wealth and still take her rank with any State in the Union in material advantages.

The area of the State is 46,810 square miles, or 29,958,400 acres, being 188 miles wide and 330 miles long, and with a river frontage along the Mississippi of 357 miles. The number of acres used as farm lands, as shown by the most recent authoritative statistics, was 17,572,547, 6,849,390 acres of which were in actual cultivation. These lands were divided into 144,318 farms, the average size of which was 122 acres. The population of the State is estimated to be 1,500,000, having increased from 791,305 in 1860.

The surface of the State generally is undulating, with a gradual slope from north to south. The Yazoo Delta is not included in this general descripton, being composed of level bottom lands and alluvial soil. The highest elevations to be found in the State are in Tippah and Union counties in the northeast, where some of the hills reach an altitude of 1,000 feet, the greatest elevation in the central portion of the State is from 300 to 500 feet, while the surface near the gulf coast is only from 20 to 30 feet above the sea level. All this part of the State is well drained by creeks and rivers

COTTON HARVEST

The Yazoo Delta or bottom lands lie in the north-western part of the State and occupy one-sixth of the area of the State, and are bisected by the main line of the Southern Railway. This section has numerous navigable streams, such as the Yazoo, Yallabusha, Tallahatchie and Sunflower rivers, and is dotted with lively and prosperous towns, like Greenwood, where the railway crosses the Yazoo River, Greenville, on the Mississippi River, the terminus of the Southern Railway, and many other smaller towns and villages. The lands of this section are among the richest in the whole world, and the region is rich in timber, among which are twelve varieties of oak, in addition to ash, locust, gum, cypress, maple, hickory, wormwood, and others.

This remarkable section, which the great Southern Railway bisects nearly in the center, deserves more than a passing notice. It is very nearly a V-shaped piece of [illegible] the point of the V beginning at the mouth of the Yazoo River, about a mile north of Vicksburg, and running nearly to the north line of the State, the Yazoo [illegible] boundary. The Delta contains [illegible] It has a good drainage, as Horn Lake, [illegible] the Delta, is 114 feet above the mouth of [illegible] There is not a single swamp in its entire [illegible] streams flow through the Delta, [illegible] network of waterways, and there are [illegible] that are traversed by steamboats of [illegible] tons burden. In addition to [illegible] bayous, which are used in [illegible] these are available for two or [illegible] the year. The soil is wholly alluvial, h[illegible] the overflow of the Mississippi River during the ages past, and now that the river has been controlled by levees so as to prevent future overflows, nearly the entire region has become available for settlement and cultivation, and newcomers are already filling the country at a rapid rate.

The Delta produces more cotton than does any other one district in the world, though less than one-fifth its area is given to that crop.

While Mississippi may have no mineral deposits of value, good building stone is found in some localities. A fair quantity of marl is abundant, and clay in many sections is well adapted to the making of brick, tile and pottery. In nearly every part of the State flowing artesian water can be had at a depth of from 300 to 600 feet. This is a great blessing to the Delta, where this pure water has considerably lessened the danger of malarial diseases.

The climate of the State is usually mild, and is not subject to extremes of heat and cold. The summers are long, but a temperature of 95 degrees is unusual. The winters are cool and agreeable, but a temperature of 70 degrees is not unusual even in January.

Mississippi, because of its excellent natural conditions, is one of the healthiest States in the country, the official statistics showing that, while the death rate in Massachusetts is 18.59 per 1,000; New York, 17.30; Pennsylvania, 14.92, and Colorado, 15.10, it is but 12.80 in Mississippi, and this is inclusive of the colored population, the average death rate of which in the entire South is 17.28 per 1,000.

Facts will show that Mississippi is one of the best-governed States in the Union. Every householder with a family is entitled to hold exempt property sufficient to support a family in comfort. Liquor selling is regulated by "local option" in the counties. Since this law went into effect, about

eight years ago, saloons have been abolished i[n] all but six or seven counties. Purity of elections is assur[ed] by the Australian ballot system. An educational and poll qualification has [elimi]nated the ignorant and vicious voter from participating in elections. Mississippi has a smaller mortgage indebtedness than any other State except three, while the public [debt]

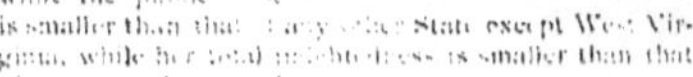

is smaller than that [of] any other State except West Virginia, while her total indebtedness is smaller than that of any other State, with [illegible] exceptions.

While Mississippi is the greatest of all the cotton-producing States, it is erroneous to presume that cotton is the only product that can be raised here. A great variety of [illegible] [have] been successfully grown. Corn, oats, hay, rye, millet, wheat, rice, potatoes, peas, sorghum, hemp and all kinds of fruit are standard crops. Stock-growing is destined to become one of the leading industries of the State. Dairying and truck-farming already yield profitable returns. Hogs and sheep are raised with great success. In verification of the agricultural worth of Mississippi the following from the pen of Mr. S. M. Tracey, late of the State Experiment Station, is offered in evidence:

"The percentage of the gross earnings of the capital invested in farms, including land, buildings, implements and stock, is very high in Mississippi, the average for the United States being 12.4 per cent.; for Ohio, 11.1 per cent.; Indiana, 10.[illegible] per cent.; Illinois 12.5 per cent.; Michigan, 12.[illegible] per cent.; Wisconsin, 12.7 per cent.; Minnesota, 17.2 per cent.; Iowa, 14.5 per cent.; Nebraska, 13.1 per cent.; Kansas, 13.5 per cent., and for Mississippi, 43.8 per cent. By this showing, money invested in Mississippi farms brings nearly three times as much as the average for the whole country, and more than twice as much as any of the States named.

"The average value of farming lands, including both improved and unimproved, is for the United States $25.[illegible]; for Ohio, $41.1[illegible]; Indiana, $12.50; Illinois $58.35; M[illegible] [illegible] 72; Wis[illegible] Minne[illegible] Iowa, [illegible] [Ne]braska, [illegible] $23.43; a[illegible] [Missis]sippi $[illegible] words, $[illegible] .0[illegible] chase [illegible] Ohio, 23.8 [in] Ind[iana], 20.7 in Illinois, 2[illegible] in Michigan, 33.3 in Wisconsin, 45.[illegible] in Minnesota, 27.7 in Iowa, 42.2 in Nebraska, 42.7 in Kansas, while it will purchase 105.[illegible] acres in Mississippi. These figures speak for themselves.

"The report of the eleventh census gives some very interesting figures in regard to crop values. According to that report, the average value of farm products per acre for the whole United States is $[illegible]; for Ohio, $7.27; Indiana, $6.27; Illinois, $7.2[illegible]; Michigan, $8.[illegible]; Wisconsin, $7.25; Minnesota, $6.[illegible]; Iowa, $6.27; Nebraska, $10.7[illegible]; Kansas, $[illegible].26; and for Mississippi, $10.70. By these figures the average crop from an acre in Mississippi is worth more than 40 per cent. above the average for the whole country, and more than 25 per cent. above that of any of the States named."

The public school system of Mississippi dates from 1871, but it has been so much improved since that time that it now ranks with the best in the Union in its thoroughness and efficiency. In proportion to taxable valuation the State perhaps expends more for education than any other State. Mississippi spends annually on her public school system more than a million dollars, and for educational purposes nearly a million and a half dollars. Besides the excellent free school system, supported by a State and county revenue of $1,413,766, and possessing property valued at $1,[illegible], there are many fine institutions of learning in the State of a public nature, in addition to more than two hundred private and denominational schools. These are the State University, the Agricultural College, the Industrial Institute and College at Columbus, the Deaf and Dumb Institute, Institute for the Blind, Negro Agricultural and Mechanical College, and State Normal School. The State University, at Oxford, was founded in 184[illegible]

[illegible]

In 18[illegible] Congress granted a township of la[illegible] to the Stat[illegible] for the purpose. It has been supported by St[illegible] appropria[illegible]tions and by the interest on the proceeds of the sale of land granted by Congress. The institution ranks high among the colleges of the country. The characte[illegible] of its [illegible] has been a guarantee of its [illegible]. Some [illegible] have [illegible] impress upon the history of the Sta[illegible] [illegible] The university at present includes [illegible] education, with a school of [illegible] literature and arts, com[illegible] [illegible]rent schools. The curriculum [illegible] Latin, Greek, German, French and E[illegible] [illegible] letters, mathematics, all the natural [illegible] philosophy, political economy. The university has an excel[illegible] [illegible] and chemical and physical [illegible]. Tuition is free to all except [illegible] justly merits the repu[illegible]

[illegible] College [illegible] Columbus [illegible] educational history of [illegible] [illegible] in 188[illegible] the wom[illegible] of the State [illegible] [illegible] from [illegible] State of [illegible] three courses of study [illegible] [illegible] advantage of the

[illegible] Agricultural and Mechanical College

was located at Starkville. The discipline here is military, and while the college was established primarily for the instruction of the youth of the State in the agricultural and mechanical arts, provision is made for instruction in both common school and collegiate courses. The education imparted here is also practical and illustrative; students are required not only to be familiar with labor, but to labor themselves, which indeed constitutes an important part of their education. The buildings are handsome, permanent and commodious; the farm embraces 1,[illegible] acres of land, [illegible] of which are under cultivation, including gardens and grounds. The farm is also well stocked with improved breeds of cattle, and with a complete outfit of the latest improved agricultural implements and farm machinery.

The Alcorn Agricultural and Mechanical College was founded in 1871 and dedicated to the education of negro youth. Instruction is given in the agricultural and mechanical arts, and the courses of study college, academic, scientific, preparatory and commercial. The college has been very successful. Tuition is free, as in the college for whites, and the State has appropriated, in addition to the interest derived from the agricultural scrip fund, all the money required for successful maintenance.

[illegible], MISS.

To those who follow agricultural pursuits, and who for any reason desire to seek new homes, Mississippi offers inducements superior to many of the other States. Her climate and soil are unsurpassed. All agricultural products can be produced in abundance. Negro labor is almost the only kind employed, and farm hands are paid from $10 to $1[illegible] per month. The United States still own[illegible] [illegible] hundred thousand acres in Mississippi, and the State [illegible] hundred thousand acres, all of which is for sale at cheap rates. The State is beginning to be recognized as a field for mills and factories. Land can be procured as cheap as in any other State. The people of the State are brave, generous, loyal and hospitable. They are proud of her glorious past, contented with her prosperous present, and justly hopeful of her splendid future.

[illegible], MISS.

The Southern Railway crosses Mississippi in almost a straight line from east to west, entering the State in Lowndes County, near Columbus, and terminating at Greenville, an important point on the Mississippi River. Between these two points are located a score of the best towns in the State, including West Point, Mhoons Valley, Cedar Bluff, Maben, Mathiston, Eupora, Grady, Townsden, Stewart, Kilmichael, Winona, Carrollton, Greenwood, Itta Bena (from which a branch runs to Webbs), Moorhead, Baird, Indianola, Elizabeth and Stoneville.

Columbus is a city of nearly 6,000 inhabitants, beautifully laid out, substantially built, with fine graveled streets, and noted for the wealth, culture, refinement and hospitality of its people. It is situated on a high and commanding bluff on the east bank of the Tombigbee River, practically at the headwaters of its successful navigation. The bluff gradually slopes to the Luxapalila, a stream almost of sufficient size and importance to be utilized for navigation, and one which furnishes within a few miles of the city unlimited water power, with volume enough to set in motion millions of spindles, lathes, etc. Columbus is located just two and a half miles above the confluence of the Luxapalila with the Tombigbee river; showing that Nature furnishes to her the prime essentials to a large manufacturing city, viz., perfect drainage, and an excellent navigable waterway to the seaboard.

Columbus has superior [illegible] advantages, including model [illegible] [illegible] than any [illegible] South [illegible] built on that [illegible]

[illegible]

and College, for the education of the white girls of the State of Mississippi in the arts and sciences, are located here. To the latter school belongs the distinction of being the first State institution ever founded for the education of women. The building occupied is a large four story brick structure, surrounded by beautiful lawns and an abundance of stately oaks, elms, etc., while between the institute and College Street are fountains and a well-kept flower garden.

Columbus has a number of highly prosperous manufacturing establishments, including a large cotton mill with 5,004 spindles and 256 looms, which has never been idle a day since it was started. It consumes 700,000 pounds of cotton annually and employs 150 hands.

The city has an opera house, a dozen fine churches, electric lights and gas works, and an abundance of pure water.

West Point, a few miles west of Columbus, has a population of 3,500. It is located in the midst of a sandy plain about four miles square, around which are the finest prairie and creek bottom farming lands in east Mississippi. This immediate section of the State has for generations been famous for the abundance of its crops. West Point is comparatively a new town, but

WINONA, MISS.

is improving with great rapidity and has several manufacturing establishments. During the past season [illegible] handled over 5 [illegible] bales [illegible] [illegible] The Mary Holmes College, a m[illegible] [illegible] Academy, and the West Point [illegible] Business College, one of the best institutions of its kind in the South.

Winona is an active town of 2,5[illegible] inhabitants, surrounded by a section noted for its magnificent timber. Oak, hickory and beech predominate in such quantities as to make this [illegible] a most desirable [illegible] for manufacturing of all [illegible] where [illegible] is used. It has already become the second largest market for hickory in the United States. It is the county seat of Montgomery County, and is located on the crest of the dividing ridge between the Big Black and the Yazoo rivers, being the highest point between Chicago and New Orleans. About [illegible] bales of cotton are compressed here annually, and it is the commercial center of a rich region. Winona has good schools and churches and many advantages, including the purest of water from artesian wells of great depth. But a short distance from the town is the celebrated Stafford well, furnishing a mineral water of great efficacy and almost national reputation.

Carrollton, which is the county seat of Carroll County, has 1,000 population and several churches, [illegible] growing colleges and good schools. Its business is chiefly dependent on the prosperous agricultural region surrounding it.

Greenwood, its next-door neighbor on the west, has a population of 2,000, five churches, several public schools and one bank. The surrounding country has an inexhaustible supply of heavy oak and cypress timber. Greenwood has a cotton and cotton-oil mill, saw mill, stave factories, ice works, brick factory and machine shops.

Indianola, in Sunflower County, is a prosperous town of fifteen hundred inhabitants. It is the county seat, and has excellent educational facilities and several industries.

Greenville, which is the westernmost terminus of the Southern Railway and one of its two gateways on the Mississippi River, is one of the State's most important and prosperous cities. In 1865 it was a mere river landing. To-day it has 10,000 population and is growing rapidly. Its business interests are almost entirely dependent on cotton, but its growth and prosperity have demonstrated that this dependence has not been misplaced. The visitor to Greenville will be impressed with the signs everywhere present of prosperity. It is the *entrepot* of the Yazoo Delta, than which no more fertile region exists. It takes pride in its court house, the finest in the State, in the stability of its banks and commercial houses, and in the purity of its water, which comes

GREENVILLE, MISS.

from a number of deep artesian wells. The excellent public school system of [illegible] is a graded one, consisting of [illegible], the last two of which form the High School, [illegible] Latin [illegible] modern branches [illegible] and a graduate finds ready admission in the colleges of the State. [illegible] and the [illegible] Among the [illegible] the great [illegible] many thousand [illegible] of mall an [illegible] Mississippi R[illegible] Greenville are [illegible] The Southern Railway, the [illegible] to

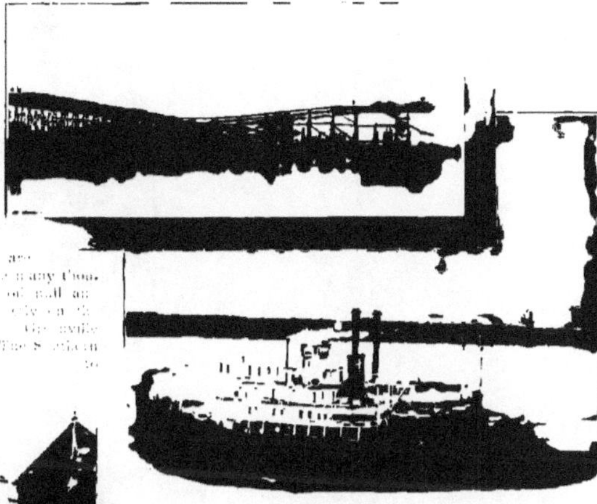

its own barge line on the river, the Alabama [illegible] coal sold for delivery at Mississippi River points south of Greenville in competition with coal [illegible] from Pennsylvania.

The Southern Railway [illegible] the State northeast of Meridian and [illegible] the New Orleans & Northeastern for New Orleans, [illegible] distant.

Meridian has been termed the electric city of Mississippi because of her remarkable growth. She has [illegible] population, and is progressive, enterprising and alert to her business opportunities. Her location is in the midst of one of the richest agricultural sections in the South, and her citizens, represented by the Young Men's Business League, have brought about her recognition as a manufacturing as well as a commercial [illegible]. The [illegible] or more prominent [illegible] employing [illegible] her of operatives [illegible] and [illegible] value of [illegible]

output is very great. The city has splendid water and gas works, an extensive electric lighting system, electric street cars, and a sewerage system which cost upward of $[illegible]. The streets are paved and kept in the best of order, and from every point of view Meridian will impress the visitor. She has 31 churches of all denominations, five modern brick school buildings well equipped, a good commercial college and two female colleges. Her banks carry deposits of nearly $[illegible], and four building and loan associations are having a prosperous existence. On the whole, Mississippi and the South may take pride in Meridian and what it has accomplished!

The Memphis division of the Southern Railway cuts across the northeast corner of Mississippi, the chief town upon the line in the State being Corinth, a place of [illegible] population. It is located in the center of a fertile agricultural region and has all the mechanisms of a place of much greater size, including electric lights, good schools, banks, etc.

A GREAT deal of romance has been written about the State of Louisiana. Its climate, its traditions, its varied customs and varied population, the naturally artistic temperament of its people, its wonderful history reaching back to the infancy of a new continent and a new epoch in the world's life, all have tended to foster this. People visiting its foremost city love to look up the home in which dashing Lafitte, the pirate, lived, to hear traditions of him, to study the architecture of bygone generations.

Rightly told, material Louisiana is a romance. Earth and forest alike cry out to keen investing instinct with promises so fair as to excite wonder; at first glance, incredulity. Capital can realize a usury of interest in many kinds of legitimate investments here, and the laboring man has as promising opportunities of owning his home as in any other spot on the American continent.

The twenty-eight million acres of soil comprised in the limits of this superb State afford opportunities for a variety of industry as striking as is the variety of the composition of its present citizenship. Time was when sugar and cotton measured the limits of its agricultural industries. Rice was later added to the list, and for a time these three constituted the State's main industries. That time is gone now, and although the [illegible] great industries will continue to champion a vast deal of attention and employ a great deal of men and money, the present epoch is marking the development of wonderful new possibilities. A great many methods are being changed, a great many new views are penetrating. There is no hazard in predicting that the time is at hand when Louisiana will cease sending her immense cotton crops to English spinners. The loom instead will [illegible] through-out her [illegible] cities.

Louisiana's southern limit is 29 degrees [illegible] minutes from the equator and it extends northward to the thirty-third degree. The orange blossom blows into the [illegible] in its southern limits, and the navigation of its northern streams is never impeded by the ice of winter, which only at rare intervals permits the song of the skate. The mighty Mississippi splits the State in two, and geologists say that once upon a time, way back in history, the mouth of the river surged into the gulf high up above New Orleans, which is now over a hundred miles from the gulf. [illegible] this may [illegible] certain it is that the land along the stretch of this river is of such [illegible] formation. The [illegible] are known as [illegible] and [illegible] not alone along the Mississippi but [illegible] the banks of all [illegible]. The uplands of the State [illegible] a man may mention [illegible] State [illegible]

[illegible]

[illegible] wonderful fecundity of that class of Louisiana lands. At no point in the State do the uplands attain a great [illegible] [illegible] the level of the sea, and [illegible] [illegible] grazing country, which now under the influence of irrigation is making such marvelous strides [illegible] such vast [illegible] rolling [illegible] than [illegible] above [illegible]

William C. S[illegible]tor of the S[illegible] experimental station [illegible] divides the State approximately into five parts — alluvial, bluff, [illegible], longleaf [illegible] prairie lands. The alluvial region lies along the Mississippi [illegible] bayous, Red River and its [illegible] bayous, [illegible] of the [illegible] No [illegible] that this region is "the most [illegible] land in the world, equaled by few and [illegible] in the world in productive capacity." [illegible] could hardly be more sweeping [illegible] one of the very highest [illegible] and yet the material fact of [illegible] year in, year out. [illegible] belt running from [illegible] near the center of [illegible] the northern portion [illegible] South about doubling [illegible] for a diversity of [illegible] agricultural hill [illegible]

[illegible]

intermixed with the washings from the hillsides, results in very fertile vales where small farmers thrive with the advantages of complete self-support, the possibility being presented of rearing all that is needed for eating, outside of the regular industry of the crop. Of course this is a part of the splendid cotton belt in that strip of the United States wherein cotton has reigned as king for years. Extensive experiments have conclusively shown that this region is also destined to produce much of the fine tobacco of the world's market before many years, and some of that grown by the farmers for their individual use has a flavor unsurpassed by the product of Cuba. The tide of that sturdy immigration which under the direction of great railroad trunk lines is now reaching the rolling plains of the South has not yet reached this section of the State in full force, although each year it is receiving an increasing quota of industrious labor. When it is adequately populated it will stand out distinctively as a section where more homes are owned by their occupants than any other portion of the State and possibly of the country, for it is essentially the country of the "small farmer," where no big capital is required in crop growing.

The pine hill regions present a great uniformity of soil. They are especially valuable as timber and grazing property. Cattle and hogs thrive in them splendidly, being protected by the forest and the hills against the winter and at all times finding ample grazing. The bottoms of this section present the arable land. This is the same chain of hill country which stretches parallel with the edge of the gulf from Georgia to Texas, varying in timber wealth. That wealth has not to this day been half realized, and

in this line of natural resource to be located  the South, and Louisiana ranks foremost among the States holding this major portion. The millions of dollars have been expended in milling in this State with past few years have been devoted almost exclusively to the sawing of pine and cypress. The day is near at hand when millions more are going to find rich rewards in turning the finest woods of the world, to be found in these forests, into the finest furniture of the world.

Men thoroughly in touch with the situation assert that the history of the cypress industry will be repeated. Only a few years ago there was, as has been said, but little activity in cypress sawing, which is a wood almost peculiar to this State, so commercially considered. To-day, the Cypress Lumber Manufacturing Association represents an output of almost [illegible] feet of finished cypress lumber per year, and this is steadily on the increase. It has come to be foremost in the woods of the State, and it is conclusively argued that the hard woods of the State and all those capable of fine finish are soon to forge to the front in the same way. Ash, oak, magnolia, beech, walnut, gums, cottonwood, maples and a number of the woods enumerated previously are found in practically limitless abundance in many regions of the State now accessible to the world by railroads, and their utilization as the basis of a great line of industry is a definite and positive matter of the future.

IN A LOUISIANA SUGAR PLANTATION

presents one of the brightest possibilities for future investment.

The region of long-leaf pine exists in the extreme eastern and western portions of the State, and seems to be a sort of variety of that character of country just described. The soil is a gray, unretentive silt, which on proper fertilization presents agricultural advantages somewhat superior to that just described. Belts of oak, dogwood, beech, etc., occur along its streams where the land is best for tillage, and rich advantages for lumbering and furniture manufacturing are held out. The manufacture of turpentine and charcoal is extensively carried on in certain portions of this section, and resin is husbanded in great quantities, and mostly shipped abroad.

The prairie region extends across the State like the hill region, parallel to the line of the gulf and of course nearer to it. It varies in surface from flat to rolling, and of late years immense possibilities are being realized in these prairie lands for the growing of rice. Irrigation has made them wonderfully productive.

Beyond the prairies toward the gulf are the marshes, an unreclaimed and possibly unreclaimable region, the perpetual heritage of the wild duck, the snipe, the plover, the pelican and the hunter.

Forestry statistics of the United States carefully compiled show sixty per cent. of the wealth of the United States

As to the extent of Louisiana's possessions in woods of value, the following computation comparing several Southern States is reliable, and may be pretty uniformly applied to all the other sorts of woods as showing the comparative wood resources. This relates to long-leaf pine alone: Alabama, 18,885,000,000 feet; Florida, 6,615,000,000; Georgia, 16,778,000,000; Louisiana 21,558,000,000; Mississippi, 17,200,000,000; North Carolina,

A LOUISIANA COTTON FIELD

229,[illegible]; South Carolina 5,[illegible]; Texas, 20,50[illegible],000.

It is seen that even the vast and imperial [com]monwealth of Texas is behind in the race of [illegible], and it is to be remembered in this connection that the Mississippi River is the great benefactor in this result. It showers its benefices alike on soil and forestry, and such mighty tributaries as the Red, the Atchafalaya and the numerous other streams and bayous with which the State is supplied have added materially to this wealth of forestry.

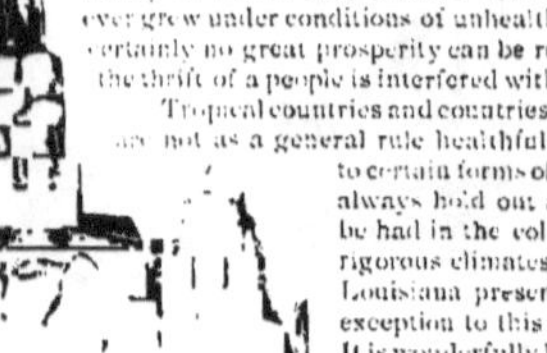

So much for the natural riches of the forests of Louisiana.

Before turning to the actual and prospective industries of the State it is well to take more than [a c]ursory notice of what [illegible] States [illegible] and in all ages has always had a most marked influence on progress, riches and civilization: watercourses. Not in the world is a similar extent of country blessed with as much navigable water as is Louisiana. There are altogether fifty-nine parishes in the State, or, as they would be called in other States, counties. Of these fifty-nine there are but four not penetrated by navigable water. When it is stated that the ag[illegible]

[illegible]

[illegible] territory in the State contain the enormous [illegible] of navigable water, upon forty rivers [illegible] of the singular kindness of [illegible]. The most important waterway, [the] Mississippi [illegible] to note [illegible] all the world, with [illegible] as do [illegible]

The matter of health is always important in the history of states and nations. No great nation ever grew under conditions of unhealthfulness, and certainly no great prosperity can be realized where the thrift of a people is interfered with by sickness.

Tropical countries and countries semi-tropical are not as a general rule healthful. Of course, to certain forms of disease they always hold out a balm not to be had in the colder and more rigorous climates of the north. Louisiana presents a striking exception to this general rule. It is wonderfully healthful, and scientific men have accounted for it by the great number of natural drains making their way to the sea within its limits. Time out of mind the Mississippi has been held to be an agency of healthfulness. Its waters, which, being at this point the aggregated drainage of over half of a mighty country, might be naturally supposed to be unhealthful, present the singular phenomenon of being the most healthful in the world. Microscopic examination reveals a singular absence of the myriad minute creations which infest almost all water, a total absence of germs, and the fact has been attributed to its swift churning current and the great abundance of fine sand or silt which permeates it. If one picks at random in four different sections of the Union any number of States, say Vermont, Tennessee, Indiana and Texas, and examines the mortality rates of the States as compiled by the general government, he will be surprised to note that Louisiana, in spite of her large area of low country, compares favorably with them all, and surpasses many. Not a single Southern State makes a better health showing. One of the best tests of health conditions is to be derived from the infant mortality rate, and an inspection of the records shows Louisiana, population considered, to be on a parity with the healthiest State of the Union.

The superb quarantine maintained at New Orleans and other Louisiana ports has resulted in absolutely keeping out of the State for over twenty years, prior to the winter of 18[illegible], the dreaded tropical epidemic, and it was then introduced from a neighboring State, but the general health conditions and regulations were so excellent that it was speedily stamped out.

Next in importance to the subject of health comes that of education in any self-governing people, where

[illegible]

popular intelligence [illegible] adequacy and enlightenment and [illegible] Educationally the State [illegible]

[illegible] Nat [illegible] S [illegible] introducing [illegible]

[illegible] college [illegible] learned a [illegible] cotton [illegible] society [illegible]

[illegible]

scientific workers, has been an object lesson in the matter. The most important department of education in the State, as in all States, however, is the common school. The length of the sessions has been steadily increasing in the various parishes from year to year, not so much in accordance with the advantages of resources for the work, for they have been about the same, but keeping pace with demand and public interest. Besides the public institutions, the State is blessed with a more than usual share of private institutions of all denominations, and it is interesting to pause here and remark that nowhere on the globe is the spirit of the constitution of this country so genuinely realized as in this State in the matter of religion. A tolerance which is as broad as true enlightenment exists between all [illegible].

No State in the Union holds out richer advantages to the home seeker than does this State. There are over a million acres of government lands yet in its confines, subject to homesteads. There are over two million acres of State lands. There are large tracts of railroad lands, not only awaiting occupation but finding

[illegible]

the roads willing and active to do all they can for the advancement and settling of these territories.

The State is dotted throughout with [illegible] New Orleans the metropolis of the State [illegible] of the South. The most important [illegible] Lake Charles, Alexandria, Shreveport, Baton Rouge, Donaldsonville, Plaquemine, Franklin, Natchitoches, Lafayette, Thibodaux, etc. Many of [illegible] by reason of [illegible]

[illegible] Orleans [illegible] not only the metropolis [illegible] State [illegible] but [illegible] the Southern [illegible] in the world, and [illegible] empire which is great not only in [illegible] agricultural wealth [illegible] the richest and most productive [illegible] world when its mellow and fertile [illegible]s [illegible]ated forests, and its undeveloped water[illegible] [illegible]ll all [h]ave been keyed up to the pitch of modern [illegible]ion an[d] development.

[Ne]w Orleans is interesting from any and every point of view. It is so unique in many things that it has an individuality and character all its own. It stands as a type of the few American cities which have not allowed the strident calls of trade to dull their melody of romance. Thus it is at once great in commercial life and activity, and bewitching in its poetic aspects. Its great exchanges, in which the transactions run into the millions, attest its influential position among the markets of the [wor]ld, while the vine-embowered villas and quaint old [stree]ts of its French quarter suggest to even the tran[sien]t visitor the ever-fascinating story of its early days.

It was founded by de Bienville, a French Canadian, [in 17]18, and from that time until 1803, when it passed

permanently into the hands of the Americans, it had a varied and romantic experience, with a frequently [illegible] of sovereignty, being under the control of the [illegible] Bourbon, the Spaniard and the easy-going Creole [illegible].

The [illegible] lies on the [ea]st bank of the Mis[sissippi] [illegible] above its mouth. The great stream [illegible], however, because it turns to the east [illegible] from the eastern side, so that [illegible] points of the compass [illegible] west [illegible] center of the city and [illegible]

[illegible] proportions. The [illegible] foreign

trade alone amounted to $116,840,021, a gain of more than $20,000,000 over the previous year. Of this amount upward of $100,000,000 were exports and about $16,000,000 imports. Cotton, of course, is the chief item in the city's trade, and of this staple it handled during the year 2,241,223 bales, against 1,911,281 bales the preceding year.

In addition to the cotton shipments, there were shipped from New Orleans during the last trade year 33,004,482 bushels of cereals, 422,178 barrels of rice, 1,133,234,546 feet of lumber and 9,433,000,000 staves, and an immense amount of miscellaneous products. It will thus be seen that in all that goes to build up a seaport New Orleans takes high rank among the greatest of American cities.

Aside from its shipping interests, New Orleans has been making, during recent years, great strides in all lines of commercial life. Its population, which now numbers 250,000, embraces a large percentage of energetic, active citizens, who are alert in all matters affecting the city's interests, and who are united in the endeavor to make the city a model municipality. This desire has led to the formation of an association known as the Progressive Union, whose membership embraces the leading citizens, and whose object is to advance the city's material interests, and to herald to the world its advantages and possibilities.

Magnificent new buildings have been erected which would be a credit to any American city; factories have sprung up and prospered on every hand; miles upon miles of new asphalt paving have been laid, and a system of sanitation introduced which has made New Orleans, according to the vital statistics of the Government, one of the healthiest cities of its size in the country.

Its public school system and its school buildings are the equals of those of any city, North or South, and it has a press which is continually leading the way in the advocacy of various projects for the upbuilding of the material commonwealth.

Never was the era of progress more thoroughly inspiring a city and a State, never have men given their attention more studiously and earnestly to the question of material development, and never have a people found themselves surrounded by more munificent advantages for material growth and general prosperity as the result of intelligent plans and earnest work.

There is no more interesting city in America from the tourist standpoint than New Orleans. Its winter climate is ideal, and its attractive features are innumerable, while its festivals and fetes add elements of pleasure which are as enjoyable as they are novel. The New

HOWARD MEMORIAL LIBRARY

Orleans Mardi Gras has grown into international fame and attracts by its novel and enjoyable features thousands of visitors from all portions of the globe. The aesthetic side of the city has been well developed, and there are many charming public parks with a score of monuments and statues. New Orleans people have learned to perfection the charm of out-of-door life, and these parks are enjoyed by the masses as in few other cities.

THE NEW ST. CHARLES HOTEL

The visitor will find among its hotels, of which there are several affording modern accommodations, a magnificent structure, the New St. Charles, reared upon the site of the famous old hostelry of the same name, and now one of the largest and most handsomely appointed public houses in America. Few cities throughout the entire country can boast of as splendid a hotel, and New Orleans is fully justified in taking the pride she does in it. It occupies an entire square in the very center of the city, and its architectural features make it an imposing as well as beautiful structure. Its cost was enormous, and money was spared neither in its construction nor furnishing.

Taken as a whole the modern New Orleans is a city in which the South and the country at large have just cause for pride. It occupies a commanding and important position in the commercial world, and without doubt is destined to achieve a notable and brilliant future.

[illegible]

FLORIDA

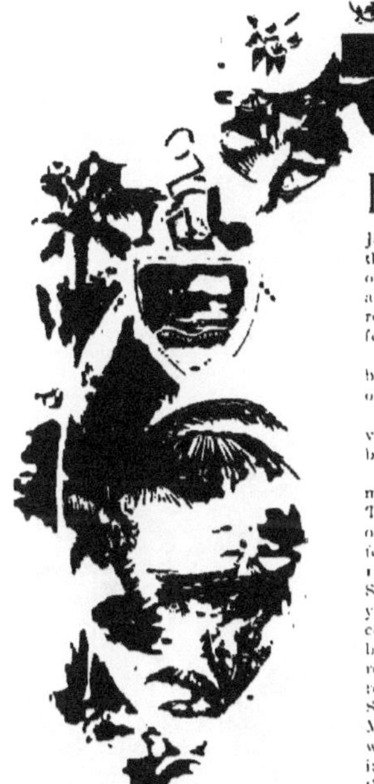

FIFTY-FIVE years before the Pilgrims set foot on North American soil at Plymouth Rock, and more than two score years before the Colonists under Gosnold, Bacon and Capt John Smith settled at Jamestown, Virginia, Ponce de Leon landed on the shores of Florida, near the site of the present city of St. Augustine. The discovery of this "land of flowers" was in 1512, and it was because of the mildness of its climate and the luxuriance of the semi-tropical foliage that the illustrious and romantic prince claimed that at last he had found the location of the fountain of eternal youth.

After a brief stay Ponce de Leon set sail to Spain, his native country, but returned again to Florida nine years afterward, only to be cruelly driven off by the natives, having suffered wounds which shortly proved fatal.

In 1520 Charles V gave one of his favorite courtiers, Pamfilo de Narvaez, an enormous land grant in Florida, and colonization was attempted, but the enterprise came to grief.

De Soto, the hardy explorer whose name is so closely associated with many of the early discoveries in the southern and western territory, entered Tampa Bay with his little fleet on May 25, 1539, giving this beautiful sheet of water the name Espiritu Santo. He evidently made no attempt to found a colony, for he was bent solely on exploration and adventure. In 1561 a band of hardy French Huguenots established themselves on the St. John's River, but were soon wiped out by sickness or desertion. Four years later the Spanish under Menendez built a fort at St. Augustine, and celebrated the occasion by swooping down upon Fort Caroline, which had been built a year earlier by the French, and massacring its inmates. In retaliation for this outrage the French sent out an expedition in 1567 and recaptured and rebuilt the fort. The following year the English admiral, Sir Francis Drake, burned St. Augustine and drove out the Spaniards. More than a century later Florida, after many vicissitudes of ownership, was given to Spain by England in exchange for the Bahama Islands, and in 1803, under what was known as the Louisiana purchase, it passed into the hands of the United States and became a separate territory by act of Congress, March 3, 1822, and took its place among the sisterhood of States March 3, 1845, being the first one added to the original and historic thirteen.

Thus the story of Florida, running back as it does to the very beginning of settlement in North America, is one in which romance, intrigue and bloodshed are dominant factors. It was the fighting ground upon which England, France and Spain conquered by turns, and wrested from

[illegible] thousand miles of steamboat navigation [illegible] F[illegible] class of seaports on the Atlantic coast [illegible] Jacksonville and St. Augustine [illegible] northeastern corner of the State, Key West [illegible] south, and Tampa, Cedar [illegible] on the Gulf of Mexico.

[illegible] coast line of Florida is [illegible] of low-lying sand islands [illegible] are called in the vernacular [illegible] number of these [illegible] estimates from ten to [illegible] end of the State a [illegible] thrown off on a tangent [illegible] and continues at [illegible] intervals for two hundred [illegible] the Key Tortugas. [illegible] Key West is one of this [illegible] midway between [illegible] and almost the [illegible] distance [illegible] welcome smiles [illegible] the

under its Venetian blue skies a charming existence amid the luxurious surroundings of costly hotels. The State has come to be the great winter playground of an ever-increasing number of people from the north, east and west, who turn toward it every fall as naturally as the birds fly southward, and who spend the entire winter there, far removed from all suggestions of ice and snow and their attendant ills.

The State topographically is divided into three almost distinct sections, and each has its devoted advocates both among the permanent settlers and the tourists who spend only the winter season.

That portion stretching along the Atlantic's shores is known both at home and elsewhere under the general title of the East Coast. It is on the main level and sandy. For a long distance it is separated from the sea by the Indian River, with its continuations, the Hillsboro and Halifax. These are rivers by courtesy only, being really salt water lakes which have in the years gone by been created by the building up of the sandbars on the shoals of the beach. These have gradually been added to until they have developed into islands varying from a few rods to a mile in width, and upon them has grown a tangle of tropical trees and vines. The oldest settlements in the State are along the

[illegible]

East Coast, and in the neighborhood of Ormond, Daytona and New S[illegible] [illegible] spots upon the Halifax, are many famous orange groves, pioneers of this branch of Florida's industry. St. Augustine is the northernmost of the resorts which have made the East Coast world famous. This spot, with the quaint relics of its old-time life, and its splendid hotels, of which the Ponce de Leon is the chief, is too well known to need more than a passing mention. A lavish expenditure of money has created here a paradise as fascinating as it is beautiful, and having such a marked individuality that it cannot be compared to any other spot in the United States.

There are pretentious tourist hotels at Ormond and Rock Ledge, and far south of the latter several famous hostelries, the Royal P[illegible]ciana and Palm Beach Inn at Palm Beach, and the Royal Palm and Biscayne at Miami. From the latter place there is a steamship line to N[illegible], N. P., and also to Key West.

The [illegible] section of the State has grown [illegible] world wide notoriety as a delightful

A FEW OF FLORIDA'S REPRESENTATIVE HOTELS

and not only add a delightful variety to the scenery but furnish opportunities for splendid horseback rides. The West [illegible] the East [illegible] [illegible] fine hotels [illegible] [illegible] West [illegible] and a [illegible] and ap[illegible] [illegible] Belleair on [illegible] The Tampa Bay Hotel at Tampa [illegible] the most magnificent [illegible] Hotel [illegible] [illegible] within the past year the Fort Myers Hotel [illegible] the [illegible] at Fort Meyers [illegible] the River. At Jacksonville there are two fine hotels, the St. [illegible] and the Windsor, both admirably appointed and [illegible] with the tourist [illegible]

In addition [illegible] the East and West Coasts and in Jacksonville, there are a large number of smaller ones scattered throughout the State, all of which enjoy a large patronage and are well kept and well known.

Of [illegible] cities of Florida Jacksonville is, by every scale of measurement, the most important. It is, from its position, the natural *[illegible]* of the State and its [illegible] and [illegible] and it has [illegible] manufacturing [illegible] miles north of Jacksonville,

A FLORIDA PINE FOREST

has the largest and deepest harbor on the East Coast, and is one of the centers of the lumber and phosphate shipping interests, a [illegible] a delightful place for re[illegible] many [illegible] homes being [illegible] here.

Palatka, fifty-five miles south of Jacksonville, is the chief city on the St. Johns River south of the latter city, and its fine hotel and many attractive features make it a popular tourist center. It enjoys also a fine business, being the natural center of a large region of [illegible] orange and fruit lan[illegible].

Tampa and Key West are each active business centers and produce [illegible] quantities of cigars. In the western portion of the State are [illegible] prominent [illegible] Pensacola, [illegible] the [illegible] and Tallahassee, the State capital, located in the heart of the beautiful hill country, where the sandy [illegible] found everywhere [illegible] of the State is supplanted by a rich loam which produces the [illegible] beautiful crops of all kinds of farm products. Tallahassee is one of the most inviting and [illegible] cities in the South. [illegible] and rich [illegible] flowers and [illegible] every season of the year. [illegible] St. James [illegible] many [illegible] of importance, including [illegible] handsome and modern

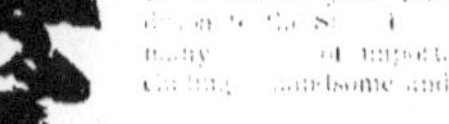

Government Building, the State Seminary and the Normal College.

[illegible] dependent [illegible] progress on the North [illegible]. While [illegible] true that many people [illegible] its great natural wealth and opportunities, consider it only as a place to go to when [illegible] winter weather [illegible] it is [illegible] making rapid strides toward wealth and improvement.

During recent years its fruit and vegetable [illegible] has grown to such proportions that more than [illegible] annually realized from the crop.

The tobacco industry has also made rapid progress and the State's production now exceeds [illegible] pounds. Under scientific culture it has been demonstrated that a tobacco which compares favorably with the finest imported varieties can be successfully grown.

Extensive investments are being made in various portions of the State by intelligent tobacco growers and

A [illegible] WHARF

each year is seeing a larger and more satisfactory crop matured. The quality of the product is steadily [illegible] proving, and results already achieved, with the favorable conditions of soil and climate, give every promise of the production at an early day of a [illegible] o [illegible] the delicate flavor of the Cuban leaf.

In the perfection of her [illegible] her wealth of picture[illegible] and tropical scenery, her splendid hotels, and her wealth of opportunity for profitable recrea[illegible] Florida offers to the tourist an ideal combination which cannot be [illegible] anywhere [illegible]. To [illegible] settler she holds out alluring inducements and bright promises, which she will do her part to fulfill.

Southern Railway
SR
AND CONNECTIONS
ATLANTIC OCEAN
GULF OF MEXICO
BAHAMA ISLANDS
WISCONSIN
PENNSYLVANIA
NEW YORK
WEST VA
KENTUCKY
TENNESSEE
MO
CHICAGO
CINCINNATI
LOUISVILLE
CHARLESTON
AUGUSTA
MACON
NORFOLK
WILMINGTON
HAVANA

INDEX

www.ingramcontent.com/pod-product-compliance
Lightning Source LLC
Chambersburg PA
CBHW022353020726
47500CB00002B/256
* 9 7 8 3 3 3 7 1 7 0 1 7 2 *